WATER

The Essence of Life

THE SUBJECTS PRESENTED IN THIS BOOK
FALL INTO THREE MAIN CATEGORIES
INDICATED BY THE FOLLOWING SYMBOLS:

 SCIENCE

 ENVIRONMENT

 PEOPLE AND SOCIETY

Water: The Essence of Life
Mark Niemeyer

Distributed in the USA and Canada by
Sterling Publishing Co., Inc.
387 Park Avenue South
New York, NY 10016-8810

This edition first published in the UK and USA in 2008 by
Duncan Baird Publishers Ltd
Sixth Floor, Castle House
75–76 Wells Street
London W1T 3QH

Credits at Duncan Baird Publishers
Managing Editor: Christopher Westhorp
Editor: James Hodgson
Managing Designer: Suzanne Tuhrim
Picture Research: Julia Brown and Gillian Glasson

Created and produced by Olo Editions
Editorial and Design: Nicolas Marçais and Philippe Marchand
Picture Editor and Layout: Emilie Greenberg
Science Consultants: Marjolaine Hamelin and Tom Benzing

Library of Congress Cataloging-in-Publication Data available

ISBN: 978-1-84483-719-9

10 9 8 7 6 5 4 3 2 1

Typeset in Din and Variable
Color reproduction by Scanhouse, Malaysia
Printed in China

For information about custom editions, special sales, premium
and corporate purchases, please contact Sterling Special Sales
Department at 800-805-5489 or specialsales@sterlingpub.com.

WATER
The Essence of Life

Mark Niemeyer

DUNCAN BAIRD PUBLISHERS

LONDON

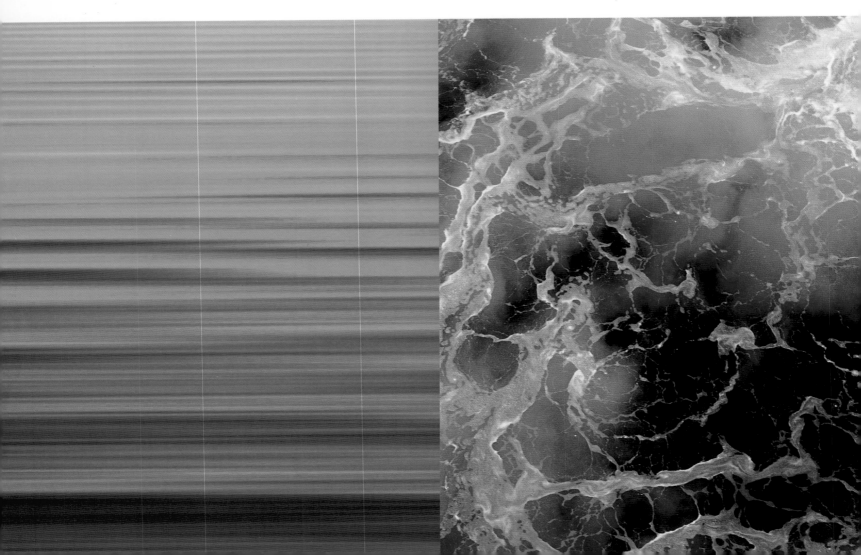

CONTENTS

Telescopes and satellites probe the universe and its bodies for the telltale signature of water, without which, life cannot exist. How fortunate we are to inhabit a planet so bountifully blessed with this precious substance. An extraterrestrial scientist, upon discovering our planet, would surely name it Water rather than Earth. Water covers 70 percent of Earth's surface, colouring it with a stunning blue when seen from space, while a gauzy veil of water enshrouds the globe as clouds.

Two atoms of hydrogen, the smallest element in the universe, combine with one of oxygen to create a molecule of water. So simple in its atomic composition, yet endowed with such strange properties that physicists are still puzzled by its behaviour. As vapour, water is a greenhouse gas, transparent to sunlight but reflecting radiant heat (infrared) back onto the surface of the planet, a phenomenon that has stabilized the temperature of the air and enabled life to evolve and flourish. Unlike most other substances the crystalline form of water (ice) is less dense than the liquid, so it floats. The infinite forms of snowflakes attest to water's ability to form a vast array of possible associations. Even when ice melts, most of the molecules remain bound together in a crystalline lattice. We are water, it makes up most of our weight. Water inflates our cells, moves oxygen, nutrients and hormones through our bodies, cools us as it evaporates from our skin and cleanses our eyes and hearts with tears.

Life began in oceans and the salt of our blood retains traces of that origin. Oceans hold 96.5 percent of Earth's water, a volume of 484,476,618 square miles (1,338,999,000 sq km). Abundant as water is, only 3 percent is freshwater most of which is frozen or underground. Sixty-nine percent of freshwater (a mere 1.7 percent of all water) is frozen in ice caps, glaciers and permanent snow and another 30 percent is groundwater, while rivers and lakes carry less than 0.30 percent. Freshwater is regenerated and redistributed by the hydrologic cycle whereby water cartwheels around the planet in an endless cycle of evaporation and condensation.

In the past century, humanity has exploded into a geological force altering the biological, physical and chemical features of the planet. Human numbers, technology, consumption and global economics have increased our ecological footprint beyond any other species. The benefits for humanity have been enormous, but a byproduct has often been the degradation of water, air, soil and biodiversity. Our relationship with water is an indicator of our ability to live in balance with the renewable necessities of the planet.

David Suzuki

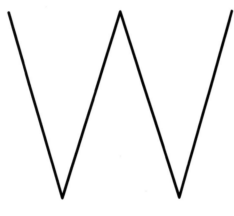ater is one of the four primary elements – along with earth, air and fire – that natural philosophers from classical times to the Renaissance thought of as making up the cosmos. While this notion seems quaint today, these early thinkers were right that water is a fundamental part of our world. Much of the Earth's surface is covered with water, and its influence on the continents and islands of the globe is enormous. Water, in its many forms, affects the weather, carves landscapes and adds to the natural beauty of our environment. "Meditation and water are wedded for ever", Herman Melville wrote,

and indeed humans seem magically drawn to the contemplation of the endless crashing or lapping of ocean waves against the shore, the pouring of a waterfall over a rocky cliff or the calm stillness of a mountain lake. Water forms an integral part of many myths and religions, as in the Old Testament book of Genesis, where God divides the waters, which existed before the act of creation itself, in order to form the heavens and the earth. And it is absolutely essential for life. Without water – this colourless, odourless substance, this elegantly simple molecule composed of two hydrogen atoms and one oxygen atom – there could be no plants or animals, not even microscopic organisms. And humans, made up of a large percentage of water themselves, could not survive.

This basic component of the planet is all around us and is constantly in motion. Throughout the hydrosphere, the entirety of the earth's aqueous envelope in all its forms, it exists naturally as a solid, a liquid and a gas, and is forever changing from one state to another. The world's water is endlessly circulating on land, under the ground, in the seas and in the atmosphere. On the surface, rivers, lakes and streams are not only in perpetual movement, but are continually changing the world of which they are a part – shifting and shaping the environment, and slowly wearing down the highest mountains. And water seeps into the ground, finding its way into every crack and crevice, sometimes creating eerie formations and underground pools in caves. In all its diversity, water is the very lifeblood of the land. Humans have always recognized the value of water, and not just as the vital liquid necessary for continued existence that they constantly need to absorb – and expel – to maintain their own delicate equilibrium. As civilizations evolved, people established villages and then entire cities near rivers, lakes and seas. Irrigation was developed early on to improve crop production, and fishing has helped feed the global population for thousands of years. Today, industry relies heavily on water, which also provides a highway for the majority of the world's shipping. The ocean, though some people may never see it during their entire lifetimes, dominates the Earth, making it a truly blue planet. It has a major influence on global climate patterns and provides some of the most dramatic examples of water's perpetual motion. The seas are also home to the largest number of the world's living organisms, some of which are so exotic or spectacular that it is hard to believe they share our planet.

In the skies, water vapour, most of which is invisible, can form clouds of impressive shapes and sizes and is important, of course, in determining daily weather conditions. As they wonder if rain or hail or sleet or snow is in the forecast, people implicitly acknowledge the importance of this water in the atmosphere and its fundamental role in the hydrologic cycle. And with good reason. The right amount of rain, just when it's needed, can provide ample supplies of drinking water and ensure abundant crops, but too much or too little can lead to flooding or drought, and widespread misery. Finally, much of the Earth's water, and indeed most of the planet's freshwater, is in the form of ice. When lakes freeze over and snow falls on a bleak winter landscape, the result can be a magical transformation – but a blizzard can mean disaster. This solid form, however, is also temporary. Frozen water may be as ephemeral as a snowflake that lands on the ground and melts almost immediately or as long lasting as the lower layers of large glaciers whose ice may remain trapped for hundreds or even thousands of years, but sooner or later it will become a liquid or gas once again.

This essential substance, present in such large quantities on the planet, is much more fragile, however, than it may seem. For if the amount of water that makes up the hydrosphere has remained more or less constant for millions of years, that does not mean it is indestructible – or at least not that its purity and the stability of its dynamic cycle can be taken for granted. Waterborne diseases, for example, have always existed, but with poor sanitation conditions in much of the world and growing populations, they put millions at risk. And while many people have plenty of water, its unequal distribution means that in some regions it is scarce and getting scarcer. Water pollution, a problem that affects virtually every corner of the globe and every aspect of life, has become a critical problem. It doesn't just poison the water we need to hydrate ourselves and to perform so many of our daily activities, but endangers all forms of life – not to mention destroying much of the world's natural beauty. And global warming, essentially a result of air pollution, is having devastating effects on glaciers all over the world, notably in the melting of the great ice sheets in Antarctica and Greenland and the ice cap at the North Pole. One of the consequences is an increasingly rapid rise in sea levels, which is particularly threatening to island and coastal populations. Taken together, all of these problems have put humankind in a dire situation. And while many measures are already being taken to reverse these negative trends, more needs to be done if a global catastrophe is to be avoided. Earth, with all its majestic splendour, is our only home. These interrelated worldwide water crises simply have to be addressed with all of the energy and imagination of which humans are capable because the entire planet depends on this vital element, which is the very essence of life.

HOLDING BACK THE TIDE

The Thames Barrier, completed in 1982 and first used the following year, is designed to protect London and the surrounding area from flooding due to exceptionally high tides. When threatening conditions are forecast, traffic on the river is stopped and the barrier is closed, preventing the temporarily rising waters – caused when a high tide is accompanied by a storm surge – from moving up as far as the capital city. Once the tide recedes, the barrier can be reopened, releasing the water that has accumulated due to the continuing flow of the river. The barrier was never closed more than twice a year throughout the 1980s, but since then it has frequently been needed more often and was used for the one-hundredth time in 2007. Though designed to provide protection until the year 2030, with worldwide sea levels rising faster than had been expected, the barrier may need to be replaced earlier than originally planned.

ON SOL

_Terra Firma

ID GROUND

1

Life evolved in water, and in all its myriad forms, life remains utterly dependent upon it. Most of the Earth's water was formed when our planet was young. It has been constantly recycled, passing in turn through trilobites, dinosaurs and mammoths, to be reused by life's next manifestations. Endless cycles of purification and redistribution not only clean water but also determine the pattern of clouds, rain, mists, lakes, rivers and wetlands that so fundamentally shape the nature of ecosystems. Life also drives water cycles. Trees form clouds and thus replenish the rain that sustains forests and the incredible diversity of life they support. From desert oases to Arctic rivers, from tropical rainforests to temperate lakes, water is life's blood. Its temperature, abundance and seasonality establishes mixtures of species and habitats. While life on Earth depends on water, one life form now shapes the nature of water more than any other. We humans

_ by **TONY JUNIPER**

EXECUTIVE DIRECTOR OF FRIENDS OF THE EARTH

have drained wetlands, diverted and dammed rivers and released contaminants into water bodies that have led to the loss of ecosystems and extinction of species. We sometimes carry on as though we are detached from the natural water cycle, as if we are not dependent on the systems that sustain the rest of life. We take water for granted. It comes in plastic bottles and pipes, and appears to be under our control. Reality is somewhat different, however. Better respect and care for water would not only help to conserve the incredible diversity of life on Earth, it would lead to a more secure future for humankind as well.

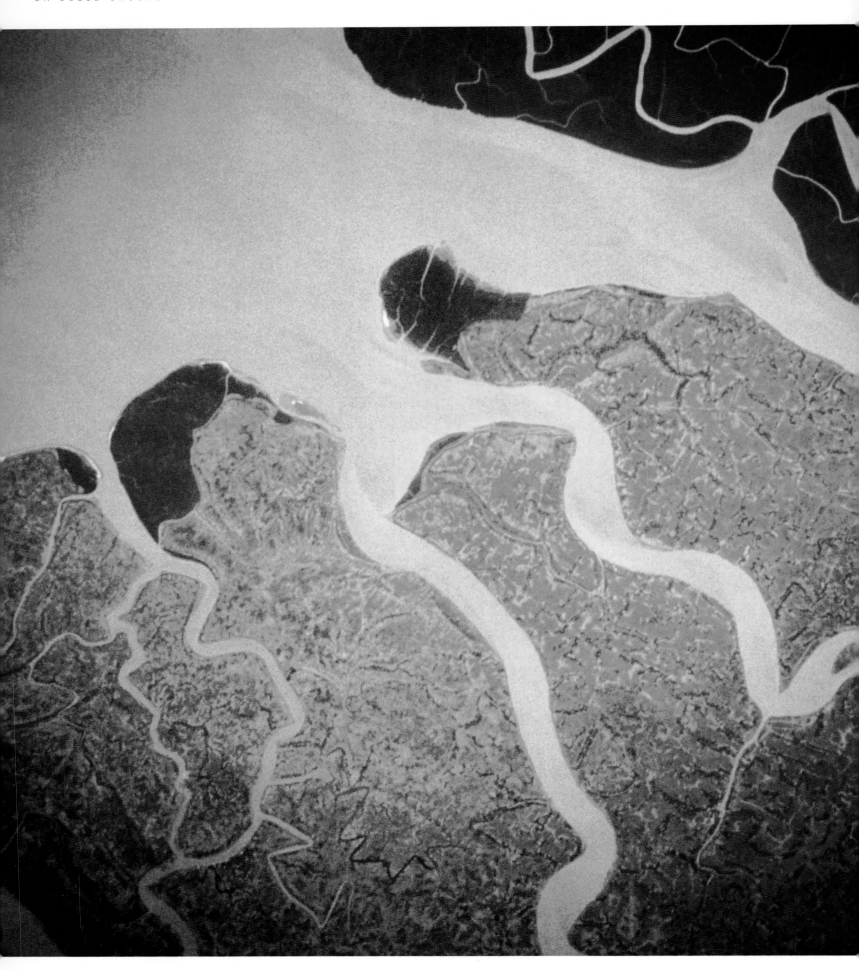

THE GANGES, one of India's holiest rivers, finds its source in the southern Himalayas. It flows through India's fertile Gangetic Plain and, after being joined by the Brahmaputra River, empties into the Bay of Bengal through the vast and rich Ganges Delta, most of which lies in Bangladesh.

SCULPTING LANDSCAPES

Water, in flowing streams, incessant waves and rains that have fallen from time immemorial, is one of the great sculptors of Earth's landscapes. In fact, this clear, simple liquid, the source of all life, can wash away tons of soil and dissolve solid rock. Mountain ranges may be built up by the powerful tectonic forces within the Earth itself, but individual hills and mountains, as well as valleys, canyons and dramatic coastlines are the result of a wearing away of the primeval building blocks of our world.

In the beginning, small surface irregularities and simple chance determine how water will flow over a given terrain. But eventually, hollows are filled, shortcuts found and softer rock worn away. Continually adjusting to the physical environment, flowing water sometimes carves out spectacular geological features. The Grand Canyon in Arizona (see illustration overleaf), which is about a mile deep, was cut into the Kaibab Plateau by the actions of the Colorado River over a period of about six million years. Overall, rivers transport an estimated 85 to 90 percent of all sediment that makes its way to the sea and accomplish most of the erosional work involved in shaping landscapes. But it's a long process. It can take a single grain of sand thousands of years before it reaches one of the world's oceans.

> "RAIN! WHOSE SOFT ARCHITECTURAL HANDS HAVE POWER TO CUT STONES, AND CHISEL TO SHAPES OF GRANDEUR THE VERY MOUNTAINS." *HENRY WARD BEECHER*

Through sedimentation, water can also work to create certain geophysical phenomena as a result of accretion rather than erosion. Flood plains and deltas, some of which can be immense, are built up by silt, clay and sand deposited by rivers. The Mississippi River Delta Basin, for example, is the result of about 5,000 years of sediment accumulation that has extended the coastline of Louisiana by up to 50 miles (80km). The basin comprises approximately 521,000 acres (211,000ha) of land and shallow estuarine water area on the edge of the Gulf of Mexico.

"EVENTUALLY, ALL THINGS MERGE INTO ONE, AND A RIVER RUNS THROUGH IT. THE RIVER WAS CUT BY THE WORLD'S GREAT FLOOD AND RUNS OVER ROCKS FROM THE BASEMENT OF TIME. ON SOME OF THE ROCKS ARE TIMELESS RAINDROPS. UNDER THE ROCKS ARE THE WORDS, AND SOME OF THE WORDS ARE THEIRS. I AM HAUNTED BY WATERS." *NORMAN MACLEAN*

THE EARTH'S ARTERIES

Rivers and streams, which form an integral part of the world's water cycle, collect surface runoff from precipitation and melting snows and carry it toward the sea. They are sometimes augmented by various underground sources and natural reservoirs, including glaciers, and often form part of a drainage network that includes lakes and waterfalls. The voyage can be a long one. A drop of rainwater that falls into Lake Superior, on the US–Canada border, for example, takes more than two centuries to travel through the interconnected Great Lakes, down the St Lawrence River and into the Atlantic Ocean. Though rivers flow through many different types of terrain, they all contribute to the formation of the Earth's varied topography, through erosion, and are the scene of diverse and extensive biological productivity. They provide habitats for myriad aquatic species as well as many land animals and birds. The Nile Delta, at the mouth of the world's longest river, hosts hundreds of thousands of migrating birds every year, including white storks, European cranes, white pelicans, and steppe buzzards.

Depending on the type and gradient of the terrain they pass through and the volume of water they carry, rivers and streams can take on a variety of forms. In mountainous regions, they often flow quickly over rocky beds and then slow down as they reach flatter land. There, they seem to take their time, sometimes meandering and creating loops, which can get cut off due to erosion and form oxbow lakes. The final destination, however, is always the sea. Rivers don't seem to take up much of the Earth's land surface, but their floodplains can stretch for miles on either side of their normal banks. Furthermore, the world's rivers make up a hydrological mosaic whose basins drain the planet's entire land surface. If the water carried by rivers represents only 1.6 percent of the Earth's surface and atmospheric water, and a tiny 0.0064 percent of the total freshwater supply, the average volume of the water in the world's rivers is nonetheless estimated at 507 cubic miles (2,115 cu. km).

THE AMAZON (LEFT), the world's second longest river, stretches 3,900 miles (6,275km) from the Andes, in Peru, to northern Brazil, where it feeds into the Atlantic Ocean. It carries a larger volume of water than any other river in the world, about 15 percent of all the water returning to the world's seas.

ICELAND (RIGHT) In Iceland's volcanic landscape, glaciers have carved out numerous valleys whose rivers and streams are fed by abundant rainfall, underground springs and the melting of winter snows. Numerous fjords and small inlets cut into the country's rocky coasts, though Iceland also possesses sandy shores.

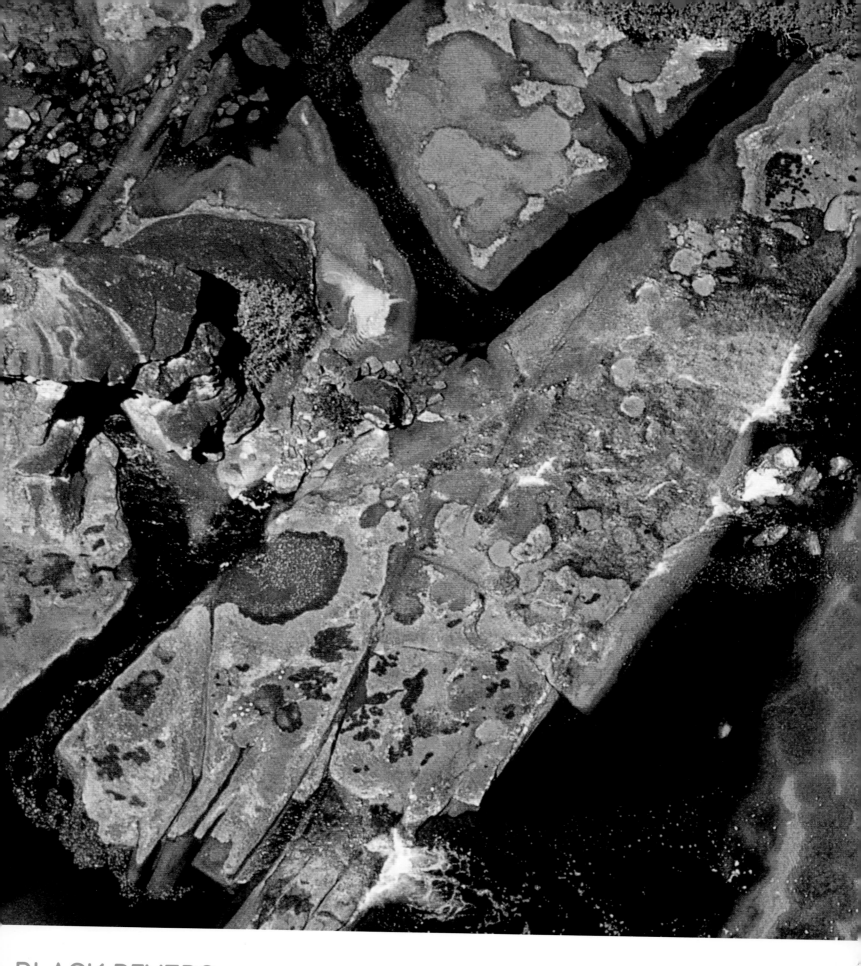

BLACK RIVERS
Sometimes rivers run black. But it's not because of any manmade pollution. Blackwater rivers, which are actually coloured somewhat like tea, are found primarily in the Amazon River system and in the south of the United States. Tannins, which leach from decaying leaves, and suspended particles of organic matter are the source of the

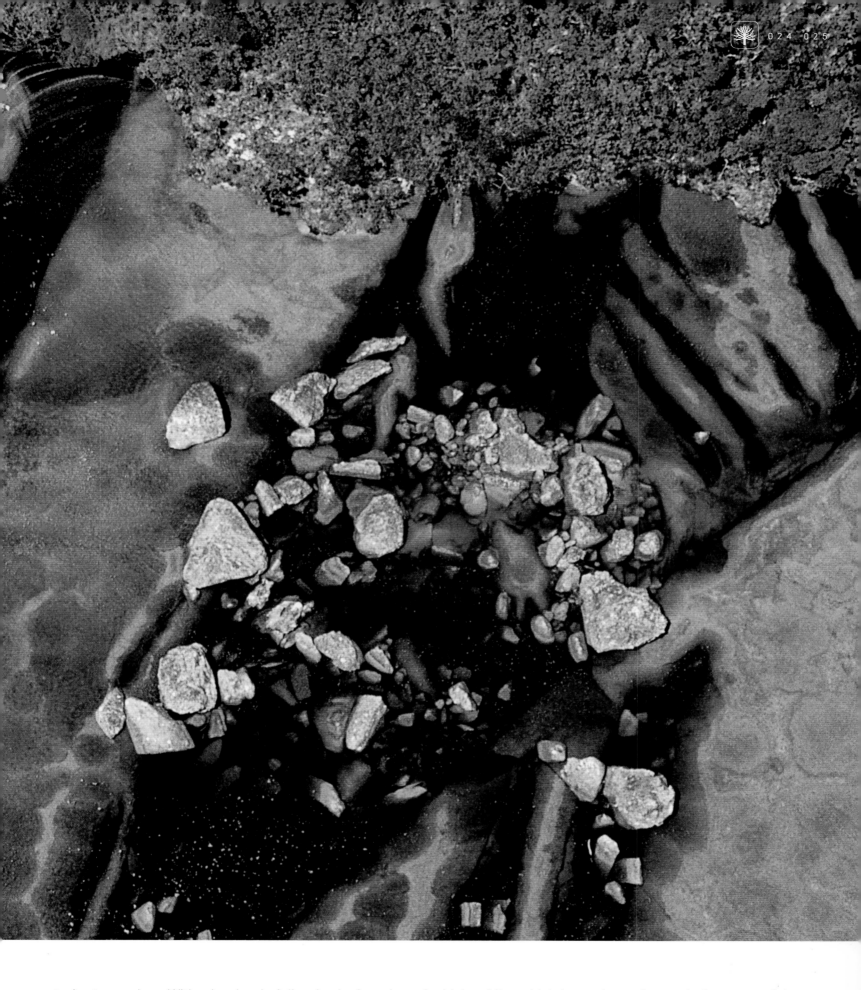

water's strange hue. With a low level of dissolved minerals and a high acidity, which keeps bacteria populations at a minimum, these rivers actually contain exceptionally clean water. This blackwater river in Venezuela, the Rio Carrao, runs through the country's Gran Sabana region, which since the 1930s has attracted prospectors looking for diamonds and gold.

ABRUPT INTERRUPTIONS IN THE COURSE OF A RIVER, WATERFALLS OR CASCADES REPRESENT AN INTERMEDIATE STEP IN THE LIFE OF A WATERWAY, BEFORE LONG-TERM EROSION FINALLY CUTS AWAY AT VERTICAL DROPS AND SMOOTHS OUT STEEP DESCENTS IN THE RIVERBED.

THUNDERING WATERS

Waterfalls provide one of the most spectacular of nature's displays. The world's highest, Salto Ángel, or Angel Falls, in Venezuela, plunges over 3,200 feet (1,114m) from Auyán-tepuí, a large plateau in the southeastern region of the country. The drop is so long that much of the water disappears into mist before ever reaching the bottom, and the falls as well as the tepuí, or flat-top mountain, are often shrouded in a veil of clouds and spray. Technically, there needs to be a vertical drop in order to create a waterfall, though the water may actually make more than one leap on its way to the bottom. A steep descent in which splashing water crashes down over rocks and boulders is more properly referred to as a cascade, and rapids are formed when water rushes and churns over a relatively shallow and rocky section of a river.

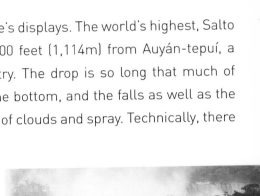

From a geological point of view, waterfalls are relatively temporary phenomena. As water pours over the lip, its erosive power eats away at the underlying rock, resulting in a sort of backward migration of the waterfall and a continuing decrease in its height. The water pounding onto the base of a fall also weakens the foundations. Every waterfall, in fact, is constantly working toward its own destruction, though new ones can be formed as a result of the uneven erosion of riverbeds, glacial movements and plate tectonics.

The second highest waterfall in the world is Tugela Falls in South Africa, at just a little over 2,000 feet (610m). But the highest waterfalls are not always the most dramatic or the most famous. A fall's volume of water, natural setting and place in history and popular culture all play a role in determining what sort of magic it will create. Niagara Falls on the US–Canada border, for example, draws about a million tourists from all over the world each year, but its height is only 167 feet (51m). And Victoria Falls, on the border between Zimbabwe and Zambia, at 343 feet (105m), is forever linked with the Scottish missionary and explorer David Livingstone, who, in 1855, was the first European to see this natural wonder and named it for Queen Victoria.

SALTO ÁNGEL (LEFT AND RIGHT), the world's highest waterfall, doesn't have anything to do with seraphs or cherubs. In fact, the falls were named in honour of the American Jimmy Angel, who, in 1937, after damaging his monoplane while landing atop Auyán-tepuí and making a heroic trek back to civilization, helped make the falls famous.

RAMONA FALLS (OVERLEAF), in the US state of Oregon, offers an enchanting display of water tumbling down a 120-foot (36.6m) stair-stepped cliff of basalt. Nestled on the well-watered western slope of the Cascade Mountain Range, it is located near Mount Hood, the state's highest peak.

"A LAKE IS THE LANDSCAPE'S MOST BEAUTIFUL FEATURE. IT IS EARTH'S EYE; LOOKING INTO WHICH THE BEHOLDER MEASURES THE DEPTH OF HIS OWN NATURE. THE FLUVIATILE TREES NEXT THE SHORE ARE THE SLENDER EYELASHES WHICH FRINGE IT, AND THE WOODED HILLS AND CLIFFS AROUND ARE ITS OVERHANGING BROWS." *HENRY DAVID THOREAU*

LAKES GREAT AND SMALL

Inland bodies of water that fill depressions in the Earth's surface, ponds and lakes hold only about 0.27 percent of the world's freshwater reserves, but they play a large role in local ecosystems and provide a source of livelihood and recreation for many people. While the most obvious difference between the two is size, lakes are usually natural formations that are linked with rivers, while ponds are often manmade. Lakes, which are deeper than ponds, also have a much greater variation in temperature between surface and bottom waters, especially in summer. Like seas and oceans, large lakes can also affect the surrounding climate, greatly increasing, for example, rainfall or winter snows.

The five Great Lakes of North America – Superior, Michigan, Huron, Erie and Ontario – alone contain about 20 percent of the world's freshwater. With a total surface area of about 94,250 square miles (244,098 sq. km), about the size of the entire United Kingdom, they form the largest group of freshwater lakes on the planet. If all the water in these lakes were spread out evenly across the continental United States, it would cover the country with almost 10 feet (3m) of water. Located on the US–Canada border (only Lake Michigan is entirely within the US), they are connected together in a chain that is drained by the St Lawrence River. Since the opening of the St Lawrence Seaway in 1959, deep-draft vessels have been able to gain passage from the Atlantic Ocean to the Great Lakes, on whose shores are located several major cities, including Buffalo and Chicago in the US and Toronto in Canada.

Africa has its own group of Great Lakes. Located primarily in the Great Rift Valley in the east central part of the continent, they are usually defined as including lakes Turkana, Albert, Victoria, Tanganyika and Malawi. Lake Victoria, lying mainly in Tanzania and Uganda but also having a border with Kenya, is the second-largest freshwater lake in the world after Lake Superior. Its existence was made known to the Western world in 1858 by the British explorer John Hanning Speck, who was searching for the source of the Nile. He correctly identified the lake – named, like the famous falls about 1,000 miles (1,600km) farther south, in honour of Queen Victoria – as one of the main reservoirs supplying the legendary African river.

MOUNT ASSINIBOINE, on the border between Alberta and British Columbia in the Canadian Rockies, rises 11,870 feet (3,618m) above sea level. The numerous shimmering lakes in the region, as well as the Assiniboine River, are fed by melting mountain snows and glaciers.

SALT LAKES
Not all lakes contain freshwater. In fact, if one defines a lake simply as a body of water surrounded by land, then the largest lake in the world is actually the salt-water Caspian Sea, between Europe and Asia. Saline lakes usually form in arid climates and occupy closed basins in which stream outflow is limited, allowing salt and other minerals to build up as

water evaporates. In fact, these inland bodies of water, which include several in South America (above, Argentina), the Great Salt Lake in Utah, and the Dead Sea between Israel and Jordan, can contain salt concentrations up to seven times greater than seawater. The banks of saline lakes are often encrusted with deposits, and can be an important source of minerals such as halite.

SAINT LEONARD, THE LARGEST UNDERGROUND LAKE IN EUROPE, WAS DISCOVERED IN 1943 INSIDE A MYSTERIOUS GYPSUM CAVE IN THE VALAIS REGION OF SWITZERLAND. THOUGH LOCAL INHABITANTS HAD LONG KNOWN ABOUT THE CAVE'S EXISTENCE, NO ONE HAD EVER BEFORE DARED TO VENTURE INSIDE.

BENEATH THE GROUND

The Earth's most abundant renewable source of readily available freshwater lies beneath the surface. Though groundwater represents only about 30 percent of this precious liquid, almost all the rest is locked up in glaciers. In fact, freshwater lakes and wetlands, rivers and streams, and soil moisture all combined make up less than one half of one percent of the global freshwater supply. In the United States, approximately a quarter of all drinking water comes from groundwater, and worldwide, about two billion people depend on this subterranean resource for their domestic needs. It is also used extensively throughout the globe for irrigation and industrial purposes.

Groundwater is an important part of the hydrologic cycle. It begins as rain or snow and seeps down through soil and rock, filling pores, cracks and other open spaces below the Earth's surface. If an underground zone becomes saturated and is permeable enough for the use of a well, it is referred to as an aquifer and can offer a source of freshwater for human consumption. The top of the saturated zone is called the water table, and its level can vary from season to season and from year to year. A drought, for example, combined with human withdrawal of water through wells can result in a significant lowering of the water table. In any case, whether it bubbles up in springs, feeds into lakes or streams, is pumped up by humans or finds its way to the ocean, groundwater eventually returns to the Earth's surface and continues the water cycle.

Technically, underground rivers, lakes or pools are not aquifers, but they can be a useful source of freshwater. Recent discoveries of large underground lakes in the Darfur region of Sudan and in Egypt, for example, should help meet some of the domestic and agricultural water needs in those areas. It is estimated that in Darfur this hidden body of water may be as large as Lake Erie and able to supply a thousand or more wells. Underground water also helps create the magnificent beauty of subterranean geological features. Stunning stalactites can form on the walls and ceilings of limestone caves through the slow dripping of mineral-rich water. Below them, stalagmites build up from the floor, and, given enough time, the two sometimes join together to create columns reminiscent of gothic cathedrals.

XPUKIL
Massive stalactites descend from the walls of a cave above a subterranean pool, near the town of Xpukil, on Mexico's riverless Yucatán Peninsula. Deep sinkholes like these, known in the region as *cenotes*, provided a source of freshwater for ancient Mayan populations.

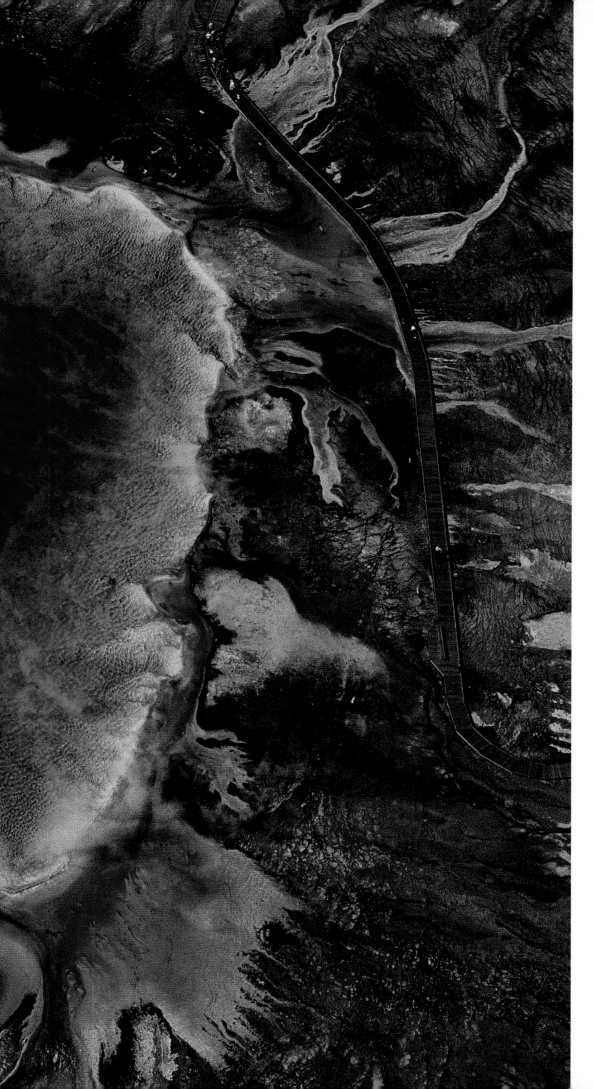

GEYSERS AND HOT SPRINGS

When underground water becomes heated through geothermal activity, it sometimes escapes to the Earth's surface. Geysers, the most dramatic example of this phenomenon, shoot out columns of hot water and steam. The most famous of these, Old Faithful, in Yellowstone National Park in Wyoming, erupts about once every hour and a half, spewing out up to 7,000 gallons (8,400 US gal, 31,800l) of boiling water in just a few short minutes to heights of up to almost 200 feet (60m). In fact, with its more than 300 geysers and thousands of hot springs, mudpots and fumaroles (or steam vents), Yellowstone contains about half of the world's hydrothermal features.

The park's Grand Prismatic Pool (left), the largest hot spring in the United States, is characterized by vivid colours produced by pigmented bacteria that thrive in the warm water. Concentrated on the spring's perimeter, they form microbial mats whose colours vary depending on the season, the temperature of the water and the amount of chlorophyll and carotenoids in their composition. The centre of the pool is usually a rich blue caused by the scattering of light in its exceptionally pure and deep water.

THE NILE During the time of the pharaohs, the Nile River was worshipped as a god, and a bureau to measure flood levels was established in order to calculate the taxes that farmers were required to pay: the higher the flood, the more the land was expected to yield – and therefore the higher the taxes.

THE HUMAN ELEMENT

Water is at the foundation of humanity's development and is part of the very soul of mankind. Early civilizations were situated near major rivers and along coastlines. In the Fertile Crescent, along the Euphrates and Tigris rivers, on the Mediterranean coast and in the Nile River Valley, humans began to cultivate crops and organize themselves into complex societies. The wealth of ancient Egypt was largely due to the richness of the Nile's floodplain and the trading opportunities offered by the river and the proximity of the sea. In Rome, water played a central role in the culture, and it is estimated that at its height about 1.3 million cubic yards (1 million cu. m) of water were distributed throughout the city each day. Aqueducts brought freshwater from springs or wells, and baths took on such a central social importance that their design and construction became a major concern of a succession of emperors. As major civilizations rose and spread around the world, their principal cities

"CIVILIZATION HAS BEEN A PERMANENT DIALOGUE BETWEEN HUMAN BEINGS AND WATER." *PAOLO LUGARI*

were almost always near water. And the fact that many religious beliefs and practices include water is a testament to its importance in how humans see and understand life.

Constant improvements in agriculture, notably the use of modern irrigation techniques, have allowed a significant increase and diversification in the food supply. And much of today's industry would be impossible without large quantities of water, which is also used in the production of energy and the transportation of raw materials and finished goods. But if we have made ingenious and efficient use of the planet's water supply to help meet both our physical needs and desire for pleasure, and have integrated this essential liquid into our spiritual life as well, human activity has also resulted in widespread pollution that poses a serious threat to the world's civilizations.

LA SERENISSIMA
Venice, on the northern Adriatic Sea, is built on an archipelago situated in a crescent-shaped lagoon. Its hundred or so separate islands are connected by bridges and canals, and the city is dominated by the Grand Canal, which snakes its way from the entrance into the lagoon to the famous Piazza San Marco. In the late Middle Ages, the

city, known as La Serenissima ("The Most Serene"), was the largest seaport in Europe and the centre of a great empire. To commemorate the importance of the waters, the republic's leader, the Doge, used to perform a symbolic wedding with the sea. Today, the very existence of this unique city is threatened by the rising oceans.

RITES AND SYMBOLS

Water has always held an important place in cultures and religions throughout the world. It symbolizes life, purity and regeneration, though it can also represent destruction. In many stories of creation, the world begins as a vast body of water from which land only later emerges. The Bible states that God created dry land by gathering together the waters, and, according to the Iroquois, the earth began as the back of a large turtle that swam to the surface of the great waters and grew to become an immense island. In Genesis, God also punishes humans by drowning the world, and many other religions include versions of a great flood story. But water is also central in many celebrations. In India, during the festival of Diwali, which for Hindus represents the beginning of a new year, lanterns are placed along the banks of rivers, such as seen here at the holy city of Varanasi, on the Ganges (left).

According to Greek mythology, the souls of the dead were ferried across the River Styx by the deity of the lower world, Charon. The river, which was supposed to flow around the infernal regions seven times, separated the world of the living from the world of the dead.

DEPENDING ON DIET, CLIMATE DIFFERENCES AND THE EFFICIENCY OF LOCAL FARMING TECHNIQUES, IT TAKES BETWEEN ABOUT 400 AND 1,000 GALLONS (480 AND 1,200 US GAL, 1,820 AND 4,550L) OF WATER DAILY IN ORDER TO PRODUCE THE FOOD TO FEED JUST ONE PERSON. THAT'S OVER A THOUSAND TIMES MORE THAN THE AVERAGE PERSON DRINKS IN A DAY.

INSATIABLE FIELDS

Humans have been cultivating crops since the Neolithic Revolution about 10,000 years ago, when the transition from hunter-gatherer communities to more settled societies based on agriculture and the domestication of animals began throughout the world. Early forms of irrigation were introduced as far back as the sixth century BCE in Mesopotamia and Egypt, where emmer wheat and barley were the main crops. In China, irrigation systems were the basis of power for early dynasties beginning around 2200BCE. Today, about 70 percent of the freshwater used on the planet is for agricultural purposes. This average, however, masks the fact that while in industrialized countries the proportion is around 30 percent, in less wealthy regions over 80 percent of water consumption goes to agriculture.

Of course, the amount of water used in agriculture also depends on the amount of rainfall, which, in fact, supplies the majority of farming needs. In England, only about 1 percent of water consumption is devoted to agricultural activity. Still, irrigation plays a key role in filling in when nature's supply is insufficient, and it is only through the combined use of irrigation and natural rainwater that most of the planet's population is able to be fed, without mentioning meeting the demand for non-food crops like cotton, rubber and industrial oils.

Agriculture, however, puts a heavy strain on the world's freshwater supply, and population growth is only going to make matters worse. Furthermore, the situation is exacerbated by both inefficient use and contamination of water, of which farming is both a cause and a victim. Much of the water used in irrigation, for example, evaporates before ever getting to the plants that need it. And pesticides and fertilizers used throughout the world eventually find their way into the groundwater, polluting water supplies of entire regions. In some countries, unclean water used in agriculture ends up contaminating crops and transmitting diseases to consumers and farm workers. The scenario is not necessarily catastrophic, however. The United Nations estimates that with more efficient agricultural techniques, it will be possible, by 2030, to increase food production by 67 percent, in comparison with levels in 2000, while restricting the corresponding rise in water consumption to just 14 percent. That goal, however, will require forethought and discipline as well as better water-management techniques, such as less wasteful irrigation methods.

RICE FIELDS (RIGHT)
Terraced rice fields on the volcanic landscape of Bali are irrigated using a network of canals. Much of the world's population, especially in East and Southeast Asia, relies on rice as a staple food, and Indonesia is the third largest producer of this vital commodity.

STRAWBERRY FIELDS (OVERLEAF)
At an industrial-scale farm in Homestead, Florida, strawberries are grown through holes in plastic sheets that cover the fields, helping to retain moisture in the soil and prevent the growth of weeds. The plants are irrigated using movable water guns.

IT TAKES ABOUT

5.5 GALLONS (6.6 US GAL, 25L)
OF WATER TO GROW
A POTATO

11 GALLONS (13.2 US GAL, 50L)
TO GROW
AN ORANGE

26 GALLONS (31.2 US GAL, 118L)
TO MAKE
A GLASS OF WINE

37 GALLONS (44.4 US GAL, 168L)
TO MAKE
A PINT OF BEER

528 GALLONS (634 US GAL, 2,400L)
TO MAKE
A HAMBURGER

IN SOME INDUSTRIES, NEW PRODUCTION TECHNIQUES HAVE VASTLY REDUCED WATER CONSUMPTION. A PAPER MILL IN FINLAND, FOR EXAMPLE, CUT THE AMOUNT OF WATER USED PER UNIT OF OUTPUT BY 90 PERCENT OVER A 20-YEAR PERIOD BY CHANGING FROM CHEMICAL TO THERMO-MECHANICAL PULP PRODUCTION AND BY INSTALLING A WASTEWATER RECYCLING PLANT.

INDUSTRY

After agriculture, industry has generally been the largest user of water resources on the planet, accounting for about 20 percent of worldwide consumption. Not surprisingly, this percentage is higher in developed countries than in less wealthy areas. Industrial uses of water include heating, cooling, washing and use as a solvent. It sometimes enters into the composition of finished products and can serve as a means of transportation as well. Major consumers of water include the chemical and petroleum sectors, mining and metal production, the wood pulp and paper industry and machine manufacturing. Vast amounts of water are also used in thermal electric power stations. And, of course, you can't make products like soft drinks, soups or medicines delivered in liquid form without water.

One reason that industry uses so much of this vital resource is that most of the surface water and groundwater it withdraws is not actually consumed in the production process. Some of it evaporates, often in the form of steam, and most of it is discharged as wastewater or effluent. Thus, whether or not a product contains water, there is actually a large quantity of "embedded water" in every manufactured item. It takes more than 2 gallons (2.4 US gal, 9.1l) of water to make a single sheet of paper, about 900 gallons (1,080 US gal, 4,090l) for a cotton T-shirt and 1,750 gallons (2,100 US gal, 7,950l) for a pair of leather shoes. It is estimated that to produce even a tiny microchip takes 7 gallons (8.4 US gal, 31.8l) of water and that the construction of an average passenger car requires 88,000 gallons (106,000 US gal, 400,000l) of water.

Despite industry's huge volume of water consumption, after a period of rapid growth between the 1960s and 1980s, worldwide usage has actually begun to level off. In Europe total industrial water consumption has been dropping since 1980, and even in Asia, which has been experiencing a period of major expansion, the withdrawal of water for industrial purposes is growing more slowly than manufacturing output. The reason for this is the implementation of production techniques that make increasingly efficient use of water. This has led, for example, to a halt in the previously ever-widening gap between industrial water withdrawal and actual consumption. But it is not all good news. Industry remains one of the main sources of pollution of the earth's lakes, rivers, oceans and groundwater.

PRODUCING ELECTRICITY
Steam escapes from the cooling towers of the Drax Power Station in North Yorkshire, the largest coal-fired power plant in the United Kingdom. The station supplies about 7 percent of the country's electricity.

WATER POWER

Hydropower is the most important source of renewable energy, currently accounting for about 20 percent of the world's electricity. Well over a hundred countries have built over 45,000 hydroelectric dams around the globe, with Canada, the United States and Brazil leading the way in this method of producing electricity. But if these dams – which also provide water for irrigation, improve river navigation and help protect against floods – produce energy without the dangers of nuclear power plants or the pollution caused by generating stations that burn fossil fuels, they nonetheless have a negative impact on the environment. Their construction leads to the loss of forests and productive farmland and decreases biodiversity. The human costs are also enormous. Tens of millions of people have been driven out of their homes or had their livelihoods and traditional lifestyles significantly disrupted by dams.

FLOATING DOWNSTREAM
Water has long been used in the transportation of logs from forests to sawmills. In more mountainous regions, logs were floated downstream in the spring, once melting snows had swelled the rivers. Agile rivermen would accompany the logs on their journey, using a pike pole to keep them from getting snagged or to

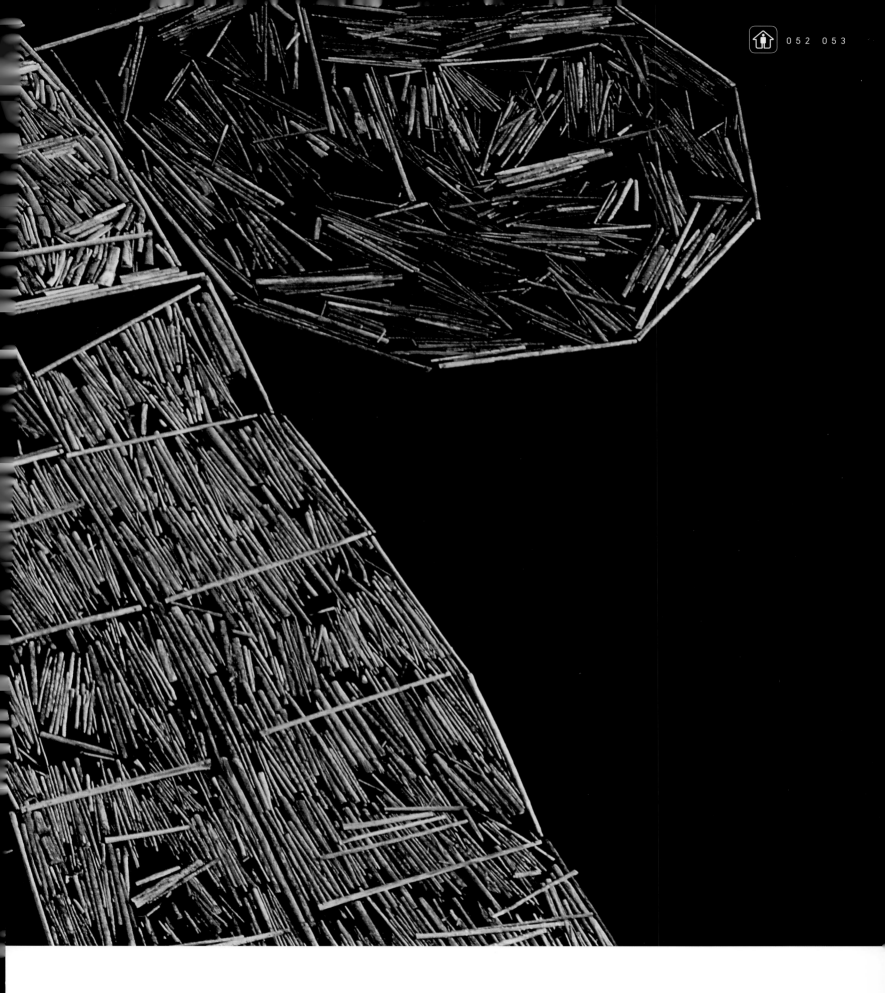

break up a logjam. In Canada (above) and the Amazon region many sawmills today are supplied with logs that make at least part of their journey by river. Since water acts as a preservative, rivers and lakes also make good storage areas while logs are waiting to be processed. Unfortunately, the rivers also facilitate illegal logging activities.

THE "POLLUTER PAYS PRINCIPLE", WHICH BEGAN TO GAIN ACCEPTANCE AFTER THE 1992 UNITED NATIONS CONFERENCE ON ENVIRONMENT AND DEVELOPMENT HELD IN RIO DE JANEIRO, MAY BE ONE WAY TO HELP MAKE INDUSTRY MORE RESPONSIBLE. ACCORDING TO THIS CODE, COMPANIES MUST ACTIVELY TRY TO PREVENT HARMING THE ENVIRONMENT AND CAN BE HELD LIABLE FOR ANY POLLUTION THEY DO CAUSE.

POISONING THE WELL

Virtually all of the world's water supplies are threatened by pollution, and agriculture, industry and homes all contribute to the problem. According to the Worldwatch Institute, toxic chemicals have contaminated groundwater on every inhabited continent, and the danger is often worst in the places where people need water most. But the world's oceans and saltwater lakes are also at risk. Since agriculture uses more water than the industrial and domestic sectors combined, it is not surprising that it contributes heavily to water pollution, primarily through the runoff of fertilizers and pesticides. In some cases this leads to the rapid growth of algae

and certain other plants, which depletes the water's oxygen level when decomposition begins. The result is not only water unfit to drink, but a disruption of aquatic ecosystems. Silt from ploughed land that is washed into streams and rivers, though not a chemical pollutant, can also damage fish spawning grounds and other costal habitats.

Industry is responsible for a whole host of pollutants. There are an estimated 12,000 toxic chemical compounds in industrial use today, and new ones are developed each year. Many of these, in varying concentrations, end up in the world's lakes, rivers and oceans. There are about 500 million tons of heavy metals, solvents and toxic sludge dumped into the global water supply each year. Another negative impact on the environment is caused by "thermal pollution" when industries discharge hot water into lakes and streams, disrupting ecosystems that are accustomed to lower temperatures.

Finally, domestic water use makes its own contribution to pollution. Every time people flush the toilet or wash their hair or do the dishes, they contribute to water pollution. Cosmetics, deodorants and pharmaceuticals all add to the lethal mix. The combined effect of all these various sources of pollution threatens the very survival of the planet. There are, however, many measures that can be taken. More natural methods of pest control, greater efficiency in industrial production and better wastewater treatment methods along with improved governance policies can all contribute to improving the situation. But time is running out.

TAILINGS (RIGHT)
are the residue left over after the minerals are removed from ore. They are often simply piled up in heaps near mines or disposed of in neighbouring ponds or rivers. Often containing various pollutants, including heavy metals, they can pose a serious threat to rivers and groundwater. Iron-mine tailings, such as these in Ishpeming, Michigan, can take on a dramatic red colouring due to the presence of iron oxide minerals.

NANHU LAKE (LEFT),
Chongqing. The environmental organization Greenpeace estimates that about 70 percent of the lakes and rivers in China are now polluted from industrial waste.

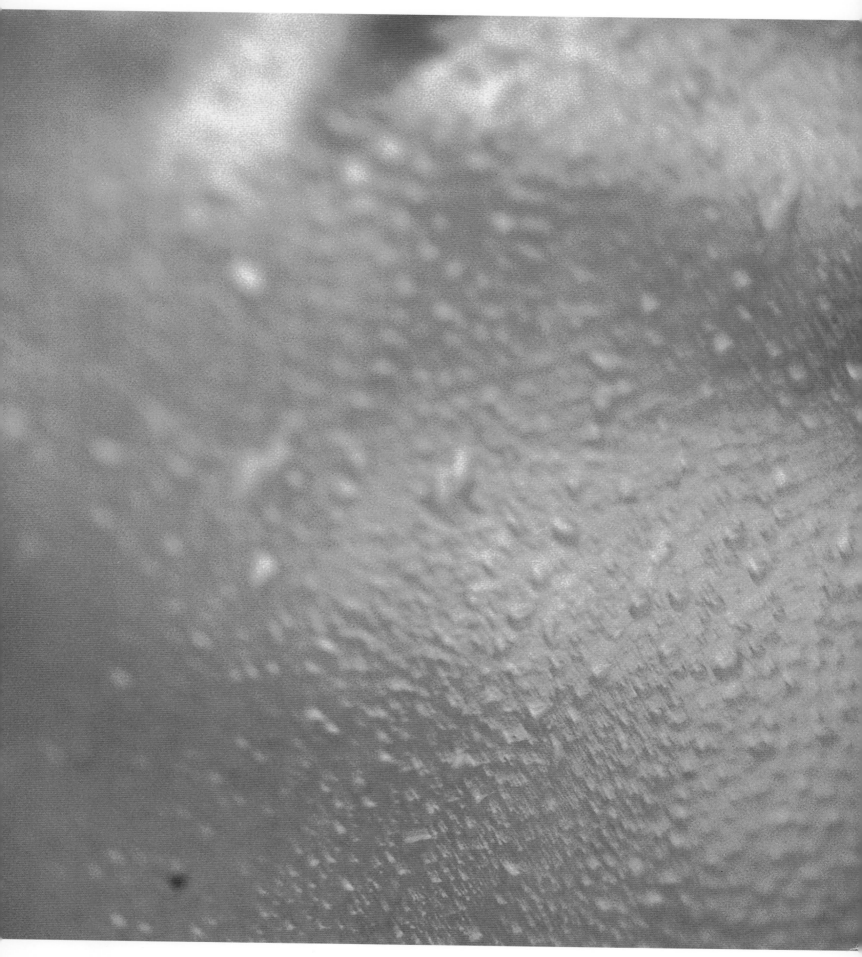

WATER CONTENT IN DIFFERENT PARTS OF THE BODY:
Lungs – 90 percent; Blood – 82 percent; Brain – 75 percent;
Body fat – 25 percent; Bones – 22 percent

THE HUMAN BODY

Almost every part of the body includes a significant amount of water in its composition. Though it has no calories and thus is not an energy source, this vital liquid is absolutely necessary for survival: a person can live for up to a month with no food, but, without water, can only last for about a week. Many important functions, like transporting oxygen and nutrients through the blood system and the body's digestive process, would be impossible without water. And it's this essential fluid that maintains, through the process of osmosis, the balance between dissolved substances, or solutes, inside and outside of cells.

The removal of wastes and toxins also requires water, which helps as well in the regulation of body temperature. Through normal elimination, sweating and simple breathing, the average person loses about half a gallon (0.6 US gal, 2.3l) of water a day, though unusual conditions like fever, vomiting or strenuous exercise – not to mention hot weather – can increase this amount significantly. Lost water must be replenished by drinking liquids or eating foods, some of which, like fruits and vegetables, can contain up to 95 percent water. The presence of water in the body also helps in cushioning joints and facilitating movement as well as in protecting organs and tissues.

But water can also be one of the greatest enemies of human health. In much of the world, safe drinking water is unavailable, transforming this liquid, so necessary to the body's normal functioning, into poison. Diarrhoeal diseases and malaria, both water-borne, pose a constant threat to many populations, and various pollutants, both manmade and naturally occurring, can transform groundwater into a health hazard.

THOUGH ALMOST ANY LIQUID WILL HELP REPLENISH THE BODY'S WATER SUPPLY, CERTAIN BEVERAGES, LIKE THOSE CONTAINING ALCOHOL OR CAFFEINE, ACTUALLY HAVE A DEHYDRATING EFFECT. AS THE BODY TRIES TO ELIMINATE THE TOXINS CONTAINED IN THESE LIQUIDS, IT EXPELS MORE WATER THAN THE DRINKS THEMSELVES CONTAINED.

DIARRHOEA AND RELATED WATER-BASED DISEASES KILL APPROXIMATELY 2.2 MILLION CHILDREN EACH YEAR, ABOUT 80 PERCENT OF THEM DURING THE FIRST TWO YEARS OF THEIR LIVES. ABOUT 42,000 DIE EACH WEEK; 6,000 EACH DAY; FOUR EVERY MINUTE; OR ONE EVERY FIFTEEN SECONDS.

WHEN WATER BECOMES AN ENEMY

While water is essential for the proper functioning of the human body, unsafe supplies, inadequate sanitation or pollution can transform it into a health hazard carrying various diseases or poisons. Every year, five million people die from water-borne diseases, and over two billion suffer from illnesses linked to contaminated water.

Many diseases, including cholera, typhoid, polio and meningitis are transmitted through water. Diarrhoea – a symptom of infection caused by a host of bacterial, viral and parasitic organisms, most of which can be spread by contaminated water – is the leading cause of death from water-related diseases in children. About 6,000 die from it each day, and in developing countries – the worst-affected areas are sub-Saharan Africa and South Asia – it accounts for over 20 percent of the deaths of children under five.

Malaria alone causes about three hundred million acute cases of illness, and over a million deaths, each year. Here too, children are the most vulnerable. Africa is by far the worst-hit area, where about 90 percent of deaths due to malaria occur. Other illnesses like Guinea worm disease and schistosomiasis (also known as bilharzias) are caused by parasitic aquatic organisms that spend part of their life cycles in water. Though usually not fatal, these diseases can result in severe health problems. The worm that causes schistosomiasis infects about two hundred million people each year, about 10 percent of which suffer severe consequences including renal failure, bladder cancer and liver fibrosis.

Chemical pollutants can also create serious health problems. While most contamination of this type is caused by industry and agricultural runoff, some poisonous substances occur naturally in groundwater. Dangerous levels of arsenic and fluoride, for example, unrelated to industrial activity, are often found in drinking water in rural regions of Bangladesh, India, East Africa and parts of China.

The answer to all these problems is increasing access to safe drinking water, improving sanitation and decreasing pollution, especially in the developing world. But the task is enormous. The United Nations World Water Development Report estimates, for example, that even if the UN goal of halving the proportion of people without access to improved sanitation is met, there will still be close to two billion people without such access in 2015.

MOSQUITOES
Some water-related diseases – including malaria, yellow fever and dengue fever – are transmitted by vectors like mosquitoes and tsetse flies that breed or live near water.

CANADA
63,382 GALLONS

UNITED STATES
46,024 GALLONS

AUSTRALIA
39,578 GALLONS

FRANCE
23,078 GALLONS

UNITED KINGDOM
7,678 GALLONS

ETHIOPIA
1,056 GALLONS

NEPAL
2,728 GALLONS

ANNUAL PER CAPITA DOMESTIC WATER CONSUMPTION
Figures for domestic use are calculated based on the amount of water supplied by
public distribution networks. They thus include water used both inside and outside the
home, as well as water withdrawn by certain industries connected to public networks.

THE THIRST FOR WATER

Worldwide, humans use about 350 billion gallons (420 billion US gal, 1,590 billion l) of water each day, and the amount is increasing faster than the population growth. Just how the world will cope with this rising demand is uncertain. Most of it goes to agricultural or industrial uses, with only about 8 percent devoted to domestic consumption. The United States, for example, uses about four-fifths of its water for irrigation and in the production of thermoelectric power, while in some parts of the world agriculture alone accounts for over 90 percent of water use.

In industrialized countries, where both the quantity and proportion of water used in the homes is the highest, only about 1 percent of the "drinking water" provided by utilities is actually consumed by people. Most of it is used for showers, baths, toilets, washing clothes and dishes, or watering lawns and gardens. A five-minute shower, for example, requires about 20 gallons (24 US gal, 90l) of water, and if you leave the tap running while you brush your teeth you will consume about another gallon and a half (1.8 US gal, 6.8l). Most people in developing countries can only dream about using water in those ways. Their domestic water consumption is approximately one-tenth the per capita rate in the West, though in some cases the disparity is actually much greater (see opposite).

All this water requires a vast production process. Water has to be abstracted from various sources, treated, stored and then finally distributed to users. The treatment process, which can include a variety of techniques, is crucial to ensuring a safe water supply. In recent years, more and more people have turned to drinking bottled water in the belief that it tastes better and is better for them. But this trend brings with it a whole new set of problems.

ABOUT TWO-THIRDS OF THE WATER USED IN THE HOME IN DEVELOPED COUNTRIES IS PIPED INTO THE BATHROOM. JUST ONE FLUSH OF A TOILET REQUIRES ABOUT AS MUCH WATER AS MOST AFRICANS HAVE FOR AN ENTIRE DAY'S WASHING, COOKING AND DRINKING.

ABOUT 40 PERCENT OF THE WORLD'S POPULATION DEPENDS ON WATER THAT FLOWS FROM A NEIGHBOURING COUNTRY.

OF THE MORE THAN 200 RIVER SYSTEMS SHARED BY TWO OR MORE COUNTRIES, SEVERAL HAVE ALREADY CAUSED INTERNATIONAL CONFLICT.

THERE ARE NOW MORE THAN 2,000 TREATIES BETWEEN COUNTRIES THAT RELATE TO WATER RIGHTS.

WATER FLUORIDATION HELPS PREVENT DENTAL CAVITIES, BUT THERE IS SOME OPPOSITION TO THE PRACTICE BY THOSE WHO FEAR IT MAY ALSO HAVE HARMFUL EFFECTS OR WHO SEE IT AS COMPULSORY MASS MEDICATION. IN THE UNITED STATES ABOUT TWO-THIRDS OF THE WATER SUPPLY IS FLUORIDATED WHILE THE FIGURE IS APPROXIMATELY 10 PERCENT FOR THE UNITED KINGDOM.

MAKING IT DRINKABLE

Less than 1 percent of the earth's water supply can be used for drinking water. Before it gets to your kitchen sink or bathroom shower, however, it has to go through a long process. First, water is abstracted from various sources. About half of all drinking water and 40 percent of the water used for industrial purposes is pumped up from underground reserves, or aquifers. Lakes and rivers, as well as manmade reservoirs fed by streams and replenished by rainfall, also contribute to the supply. One new technology known as atmospheric water generation uses a cooling process to condense and extract water vapour from the air.

After abstraction, the water needs to be treated. In general, this occurs near the area where it will finally be used, in order to reduce the cost of delivery and minimize the possibility of the water becoming contaminated. While methods vary, typically water will go through screening, sedimentation, filtration and disinfection before being safe to drink. Screening removes any large debris like leaves or trash, while sedimentation involves allowing the water to flow slowly through large tanks or basins so that suspended particles can settle out. Filtration, usually through sand with a layer of carbon or coal, takes out most of the remaining particles and organic compounds. New techniques of ultrafiltration, however, use membranes with microscopic pores to eliminate particles and microorganisms. Finally the water needs to be disinfected. While the most common method is chlorination, other techniques like ultraviolet radiation are sometimes used.

Treated water is next placed in underground or ground-level tanks, usually made of concrete or steel and often used for intermediate storage, or in elevated tanks or water towers, from which it can be distributed directly. The pressure needed to deliver the water is usually furnished by pumps or gravity (as in the case of elevated holding tanks), though some smaller water systems use air pressure. Distribution systems, which deliver drinking water to individual households, represent a major capital investment. Comprising extensive networks, they form an important part of a city's and a nation's infrastructure. In the United States and Canada alone, approximately one million miles of pipelines and aqueducts are used to carry water to and from treatment plants. That's enough to circle the earth 40 times.

WATER TOWERS (LEFT), brightly painted with cabana-style blue-and-white striping, constitute one of several groups of flat-topped water towers outside Kuwait City that hold reserves of the country's precious water supply.

WATER TREATMENT (OVERLEAF) Floating ventilation turbines in a wastewater treatment lagoon north of Baton Rouge, Louisiana, spray out polluted sludge taken from the bottom of the basin. This process of aeration stimulates the growth of protozoa and bacteria that help decompose organic matter in the water.

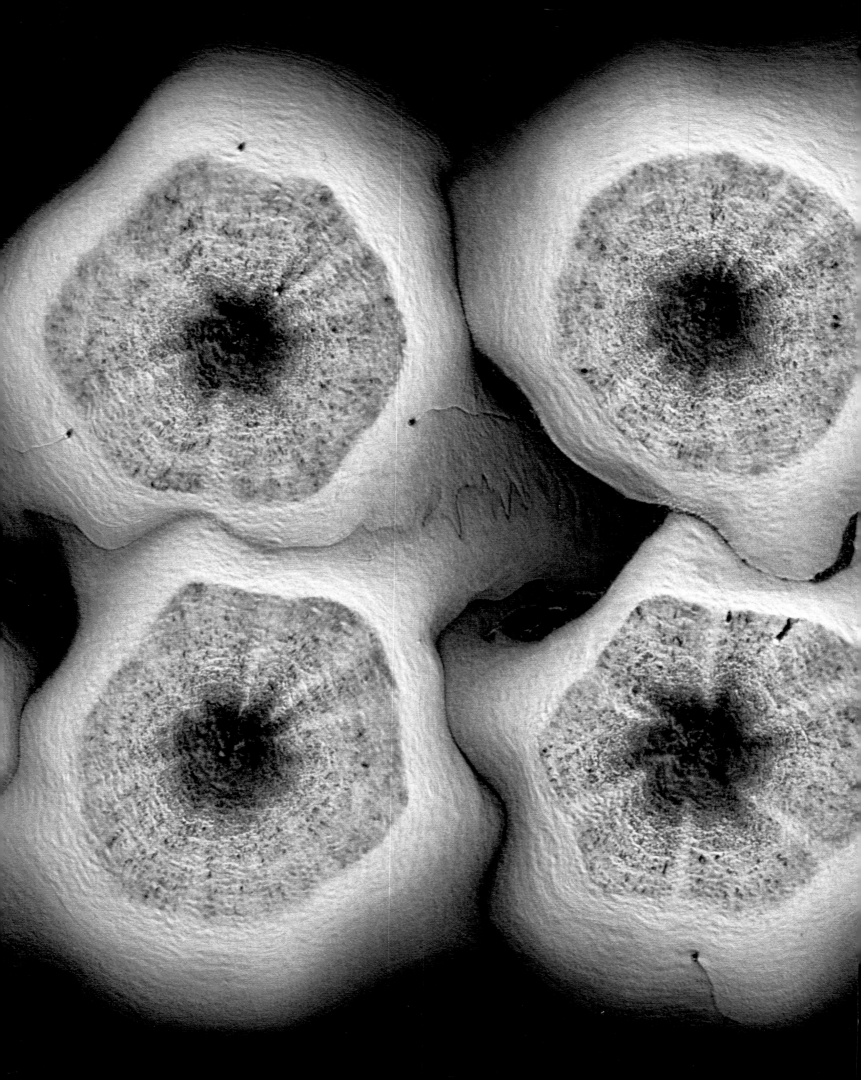

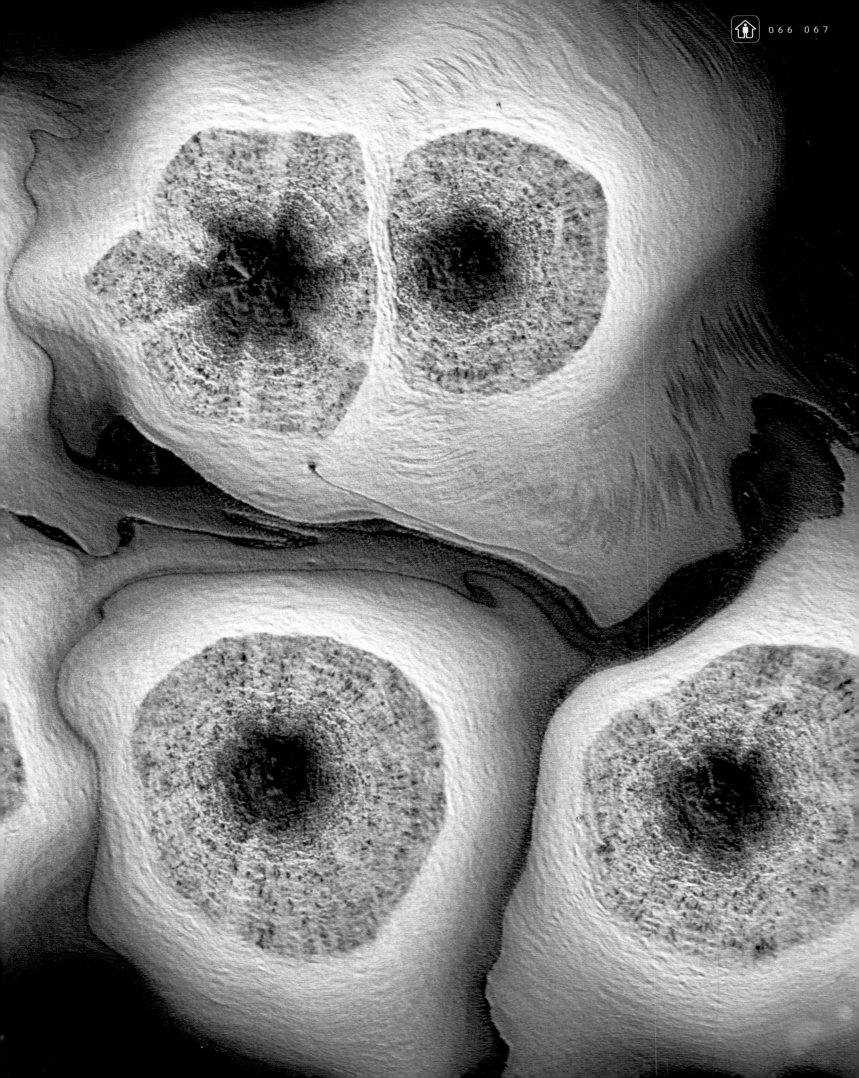

THE GLOBAL BOTTLED WATER MARKET
IS ESTIMATED TO HAVE AN ANNUAL VOLUME
OF MORE THAN
30 BILLION GALLONS
(36 BILLION US GAL, 135 BILLION L)
AND IS GROWING BY 12 PERCENT A YEAR.

**ALMOST HALF OF GLOBAL
BOTTLED WATER CONSUMPTION
TAKES PLACE IN WESTERN EUROPE,**
WHERE THE PUBLIC ALREADY HAS
COMPLETE ACCESS TO
SAFE DRINKING WATER.

ONLY 40 PERCENT OF BOTTLED WATER
IS SPRING OR MINERAL WATER,
THE REST IS "PURIFIED",
YET BOTTLED WATER CAN BE UP TO
10,000 TIMES
MORE EXPENSIVE THAN TAP WATER,
IF ONE TAKES INTO ACCOUNT THE ENERGY NEEDED
FOR BOTTLING, TRANSPORTATION
AND POTENTIAL RECYCLING OF THE CONTENTS.

NORTH AMERICA

SOUTH AMERICA

EUROPE

AFRICA

ASIA

AUSTRALIA AND OCEANIA

///// PRECIPITATION
===== EVAPORATION
▬▬▬ RUNOFF

LOST IN THE ATMOSPHERE Some regions receive more water than others. But precipitation is not the only factor in the equation. Water that evaporates is lost – only runoff , which feeds lakes, rivers and groundwater reserves, can be used by humans.

THE HAVES AND HAVE NOTS

The world's freshwater supply is not distributed evenly. While some regions are well supplied with this vital commodity, people who live in arid or semi-arid regions such as North Africa or the Middle East often face serious shortages. In fact, 60 percent of the available water – in the form of lakes, rivers, groundwater and precipitation – is possessed by just nine countries. Brazil, which has more freshwater than any other country, boasts an annual 1,975 cubic miles (8,233 cu. km) of renewable water resources. In second place is the Russian Federation, with a supply about half as large. But even in countries with an abundance of water, local variations can be considerable.

If resources vary greatly from one country to the next, what's really important is the amount of water available per inhabitant. People living in much of the world, in fact, have adequate supplies, and in places such as the Congo, French Guiana, Greenland, Iceland and Suriname there is more than 130,000 cubic yards

GREENLAND, WITH A POPULATION OF ONLY ABOUT 57,000, HAS A FAR GREATER INDEX OF RENEWABLE WATER RESOURCES PER CAPITA THAN ANY OTHER REGION. KUWAIT, MUCH OF WHICH IS DESERT, HAS THE LOWEST.

(100,000 cu. m) of available water per inhabitant per year. In contrast, places like Bahrain, Jordan, Kuwait, and Singapore all have less than 260 cubic yards (200 cu. m). Besides natural inequalities and population differences, a country's infrastructure, water governance policies and control of sanitation and pollution can greatly affect its people's access to freshwater.

Already, about a quarter of the world's population is without safe drinking water. Growing populations and increased urbanization, as well as continuing problems of pollution and climate change are all putting more and more strain on limited resources. The United Nations has reported that for many scientists a worldwide water shortage is, along with global warming, one of the two most worrying problems of the new millennium and that, if current trends continue, two out of every three people on the planet will live in water-stressed conditions by the year 2025.

DRY Australia is the driest populated continent on Earth: 70 percent of it is desert or semi-desert. The annual flow of the Murray River – source of 40 percent of Adelaide's drinking water – to the sea is now around one-fifth of what it was in 1901 and the frequency of no-flow at the river's mouth has gone from one year in 20 to one year in two.

>

WET
Australians are among the world's greatest consumers of water per person, averaging 26,400 gallons (22,000 US gal, 100,000l) of fresh-water per year. Once food production and industrial use are taken into account, they are responsible for using a total of over 5 trillion gallons (6 trillion US gal, 24 trillion l) of water per year – enough to fill Sydney Harbour 48 times over.

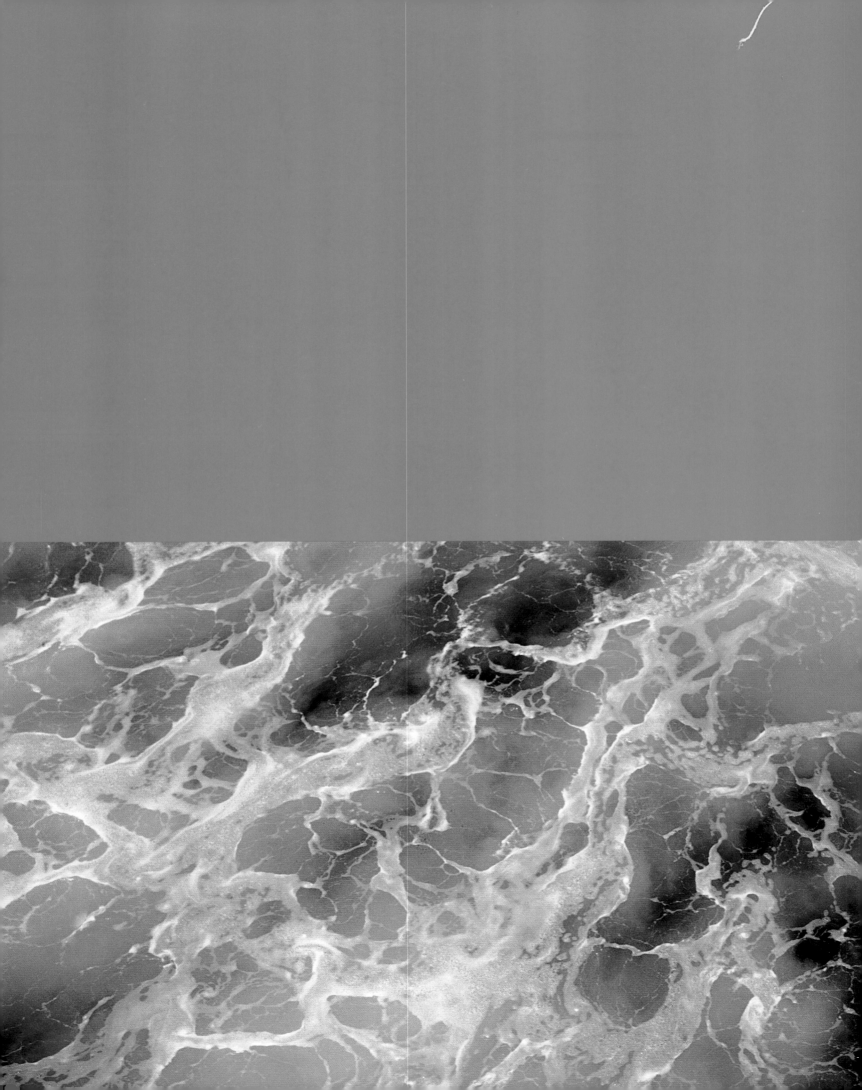

_Mare Mundus

THE SEAS

2

To most humans the sea is strange, menacing and alien. It covers three-quarters of our planet but we know less about it than we know of the surface of the moon – in fact, fewer people have descended into the Pacific Ocean's Marianas Trench than have stood on the moon's surface. We cannot be certain what lurks within the oceans' depths, but an octopus with an arm spread of 200 feet (60m) was washed ashore in Florida 100 years ago. Recent research has shown that waves in excess of 80 feet (25m) are more frequent than expected, except that is to seamen who have experienced them for centuries. Shores are still shaped by the constant battering of the sea, and tides still follow the moon, creating currents that present huge amounts of untapped energy. For those living on an island, the only means of communication with the rest of the world

_ by **SIR ROBIN KNOX-JOHNSTON**

THE FIRST YACHTSMAN TO SAIL SOLO AND NON-STOP AROUND THE WORLD

has long been by sea. Even today 95 percent of all goods moving around the globe travel by sea. From the Bible through to Melville, Conrad and beyond, the seas have inspired a rich literary heritage. So it remains no wonder that the sea still plays such a large part in our lives, practically, culturally and romantically, whether it be people from Europe, North America or Polynesia. Some arrogantly claim to have conquered the sea, but it does not even notice us when we trespass into it, the moment our wake has gone, we are forgotten. It is sometimes called treacherous, but only by those who do not know it. It sets and obeys its own rules and if we dare to venture into its territory, then we comply with those rules or we may not survive.

THE BLUE PLANET

Oceans cover almost three-quarters of the Earth's surface and contain 97 percent of the world's water. The Pacific alone, with an area of over 155 million square miles (400 million sq. km), stretches over nearly a third of the planet, but – since all the seas are connected – the Earth is really enveloped in one huge body of water, a great ocean with many basins. All this water, which conceals some of the highest mountains, deepest valleys and broadest plains on the planet, is the reason why the Earth looks blue when seen from outer space. And the world's oceans, the place where life began, play an important role in regulating the climate and weather conditions and are the source of most of our oxygen.

The oceans, of course, do not contain freshwater. Precipitation and, more importantly, the rivers and streams that empty into the sea bring with them the dissolved salts like sodium chloride that they pick up along the way. When water evaporates, the salts are left behind, which is why the oceans, like the world's saline lakes, are salty and getting saltier all the time. But there's not just salt in seawater – it contains vast quantities of other minerals as well. For example, it is estimated that the Earth's oceans contain about 25 billion ounces (714 million kg) of gold, though the concentration is so small that it's not worth trying to extract. More immediately useful are the many undersea oil reserves, and for thousands of years, trade has also been facilitated by the navigation of the seas. Fish, crustaceans and molluscs caught in the sea have been an important part of our diet since prehistoric times, and, as has been becoming clearer and clearer in recent years, the very survival of the human race may depend on preserving our oceans.

"ONLY FROM SPACE CAN YOU SEE THAT OUR PLANET SHOULD NOT BE CALLED 'EARTH', BUT RATHER 'WATER', WITH SPECKLIKE ISLANDS OF DRYNESS ON WHICH, PEOPLE, ANIMALS, AND BIRDS SURPRISINGLY FIND A PLACE TO LIVE." *OLEG MAKAROV*

THE COLOURS OF SEAS
Various shades of blue or green are what we most often see when we look at the ocean. These colours are caused by different phenomena. During the day, the sea reflects the sky and thus seems to take on some of its bluish or, on a cloudy day, greyish hue. Water's absorption of the reddish end of the light spectrum

also contributes to the colour, since it leaves primarily the blue and green colour range visible to the human eye. Microscopic phytoplankton, which contain chlorophyll, can add to this greenish effect caused by light absorption, but the phytoplankton can also appear reddish-brown at times, as in the case of the seasonal blooms in the Red Sea.

TEMPESTUOUS COASTS The constant movement of the sea shapes the coastline of Earth's landmasses. This stormy scene is at Vik, on the most southerly point of Iceland.

PERPETUAL MOTION

The ocean's waters are constantly moving. Crashing waves and the daily rise and fall of the tides are the most obvious demonstrations of this perpetual motion. But driven by the forces of the winds, the heat of the sun and the Earth's rotation, currents are what carry the huge masses of ocean water around the globe in a machine that never stops. Coastal currents move along the edge of the continents, affected by wind, waves and the configuration of landmasses. The very fact that waves do not usually hit the land head on, for example, but arrive at an angle, can create longshore currents up and down a coast.

On the open ocean, surface currents are caused mainly by wind. The continual transfer of warm air from the tropics toward the poles and the Coriolis effect – by which the Earth's rotation deflects the path of global air movements – create prevailing winds, such as the Westerlies and the Trade Winds, which drive major currents. The Gulf Stream is one of the strongest. Originating in the Gulf of Mexico, this current moves north, past the east coast of the United States and across the Atlantic Ocean toward north-western Europe. Without the warmth carried along by the Gulf Stream, places like the UK, Ireland and France would be as cold as Canada in the winter.

Deep ocean currents are pushed along by differences in water density due to variations in temperature and salinity. This process, known as thermohaline circulation, creates a global conveyer belt that constantly moves colder, saltier water from north to south – and warmer, less salty, water from south to north. This flow helps replenish the nutrients and carbon dioxide in surface waters which are necessary for the growth of algae and seaweed, which form the base of the marine food chain.

OCEAN CURRENTS, WHICH ARE ALL INTERCONNECTED IN ONE GREAT CIRCULATION SYSTEM, INFLUENCE THE EARTH'S CLIMATE AND HELP STABILIZE GLOBAL TEMPERATURES. THEY RECHARGE SURFACE WATERS WITH NUTRIENTS AND DISTRIBUTE FOOD TO BOTH MARINE AND LAND ANIMALS.

WAVES

Winds blowing across the surface of the ocean are the main cause of waves, though other influences, like tidal activity, can also contribute to their creation. The stronger the wind and the longer it continues, the larger the waves. However, the fetch area – the amount of uninterrupted ocean surface over which the wind blows and thus can exert its force on the water – and the original state of the sea before a particular wind begins to blow are also important. Though waves progress across the surface of the ocean, the water itself moves very little and does not actually follow the wave's advancing form. As it enters shallow water, a wave slows down, but its height increases. Its base begins to drag against the sea floor, and it effectively begins to lean forward, finally collapsing on itself and forming a breaker. During storms, 20-foot (6m) waves are not uncommon, and tsunamis, caused by earthquakes or underwater volcanic eruptions, can build to the height of a ten-storey building as they approach shore.

THE TIDES ARE THE EARTH'S CLOCK, RISING AND FALLING WITH AN INFALLIBLE REGULARITY. WITH THEIR USUALLY DECEPTIVELY SLOW MOVEMENT, THEY CONTAIN A POWERFUL AND IRRESISTIBLE FORCE THAT NEVER STOPS. TIME AND TIDE WAIT FOR NO MAN.

THE RELENTLESS TIDES

The moon reigns over the rising and falling of the ocean tides. The sun also has an effect on sea levels, but because the moon is about 390 times closer to the Earth its influence dominates. As the Earth turns on its axis, the moon's gravitational field, combined with the inertia of the mass of water making up the oceans, causes them to bulge out ever so slightly on the surfaces of the globe directly facing and directly opposite the moon. This small swelling creates great movements in the Earth's waters. Two high tides and two low tides occur every day, though at different times in each place and shifting a little each day, since the average interval between two consecutive high tides or low tides is about twelve hours and twenty-five minutes. Actual tidal patterns, however, are also affected by variables such as land configurations and water depths and thus tend to be rather complex. Some areas, such as the Gulf of Mexico, experience only one high and low tide per day.

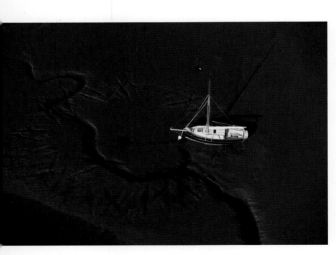

The sun's influence acts like that of the moon, and when the two bodies are on the same side or exact opposite sides of the Earth (during periods of new or full moons), combining their powers, tides reach their greatest ranges. These periods are known as spring tides. When the sun and the moon are at right angles in relation to the Earth, their gravitational forces tend to cancel each other out and create neap tides, whose differential is smaller. The variance in the height of the water between high and low tides is called the tidal range, and this difference tends to be greater along the coasts. It can also be magnified by the shape of bays and estuaries. For example, the rapid flowing tides in the Bay of Fundy, located between the Canadian provinces of New Brunswick and Nova Scotia, produce a tidal range of up to 70 feet (21m), the greatest anywhere in the world. As tides rise and fall near land, they also frequently create a horizontal movement of water pulling back and forth from the shore known as flood and ebb currents. These are the only type of currents that occur on a regular and predictable basis, and in narrow straits and inlets they can become quite swift and dangerous.

LOW AND DRY (LEFT)
When tides ebb along certain coasts or in harbours where the tidal range is relatively large, they can leave anchored boats lying helpless.

RIPPLE MARKS (RIGHT)
At low tide, the ripple action of waves is revealed in the moist sands along the shore, but just a few hours later they will be covered with water again.

TIDAL ISLANDS

Small pieces of land joined to the coast by a causeway that at high tide is submerged, tidal islands have long been attractive for their difficulty of access, making them popular as fortresses or places of religious retreat. One of the most famous is Mont St Michel, a rocky islet just off the coast of Normandy in France. This magical setting is the site of a Benedictine abbey whose origins date back to the eighth century. The difference between high and low tide in the surrounding bay can reach 40 feet (12m), the greatest in France, and the sea bed is sloped so gently that when the water recedes it exposes up to nine miles (14km) of sandy terrain. Mont St Michel, however, is no longer completely cut off from the mainland, even at high tide. A permanent causeway was constructed in the 1870s, and the bay has tended to silt up over the years – though efforts are underway to reverse that trend.

BIODIVERSITY

All life originated in the world's oceans, and most of it is still there. About 80 percent of all living things and more than 90 percent of the Earth's biomass inhabit the sea. The diversity of species is phenomenal, from tiny microbes including bacteria and algae – at the very bottom of the food pyramid – to blue whales, the largest animal ever to have lived on the planet, which can reach up to 200 tons in weight and 100 feet (30.5m) in length. But even the largest marine animals don't get as big as kelp, a type of seaweed that can grow to over 200 feet (61m) in length. There's also much more room in the ocean than on land. Since it's three-dimensional, it constitutes a huge living space with a great range of habitats that vary widely depending on depth, temperature, salinity and a host of other factors.

Life in the sea has adapted to these diverse environments in often unusual ways. Sponges, for example, have no specialized tissue; they simply filter seawater through pores in order to digest detritus and plankton. And the candy-stripe sea cucumber can defend itself by shooting sticky poisonous tubular threads at attacking fish. Coral reefs provide an environment extremely favourable to biodiversity. In fact, approximately a quarter of all known marine species find their homes in this type of habitat. The Great Barrier Reef off the northeastern coast of Australia, which extends for about 1,250 miles (2,000km) and covers an area of approximately 135,000 square miles (350,000 sq. km), is the largest reef system in the world. Formed over millions of years, it supports hundreds of species of coral (right, alcyonarian coral and thorny seahorses) as well as anemones, sponges, lobsters, crabs, prawns and about 1,600 varieties of fish and 4,000 kinds of mollusc.

MARINE MAMMALS, INCLUDING DOLPHINS AND WHALES, REVERSED THE EVOLUTIONARY DIRECTION THAT LED TO THE DEVELOPMENT OF LAND-DWELLING ANIMALS. REACQUIRING ADAPTATIONS TO LIFE IN THE WATER, SUCH AS STREAMLINED BODIES AND THE ABILITY TO HOLD THEIR BREATH UNDERWATER FOR LONG PERIODS, THEY FINALLY RETURNED TO THE SEA, LEAVING THEIR NOW EXTINCT ANCESTORS BEHIND.

A MULTITUDE OF LIFE FORMS

The sea is full of organisms ranging from the simple to the complex, many of striking beauty, which have adapted to a wide range of ecosystems. Off the shore of Nosy Be, near Madagascar, coral formations (right, top) build upward from the island's volcanic rock. Krill are small shrimp-like creatures (right, middle) that often appear in vast swarms and are considered a keystone species in the world's oceans. They feed on microorganisms and in turn become food for larger animals including fish and baleen whales. Sea turtles are well adapted to aquatic life, the development of flippers enabling them to swim more efficiently. The Hawksbill sea turtle (right, bottom) feeds on both plants and marine animals. Some species have formed a symbiotic relationship. For example, the brightly coloured clownfish (opposite page) attracts other fish toward the sea anemone's poisonous tentacles, to which it is immune. When the victim is killed, the anemone receives nourishment, and the clownfish shares in the leftovers, benefiting all the while from the protection of the anemone.

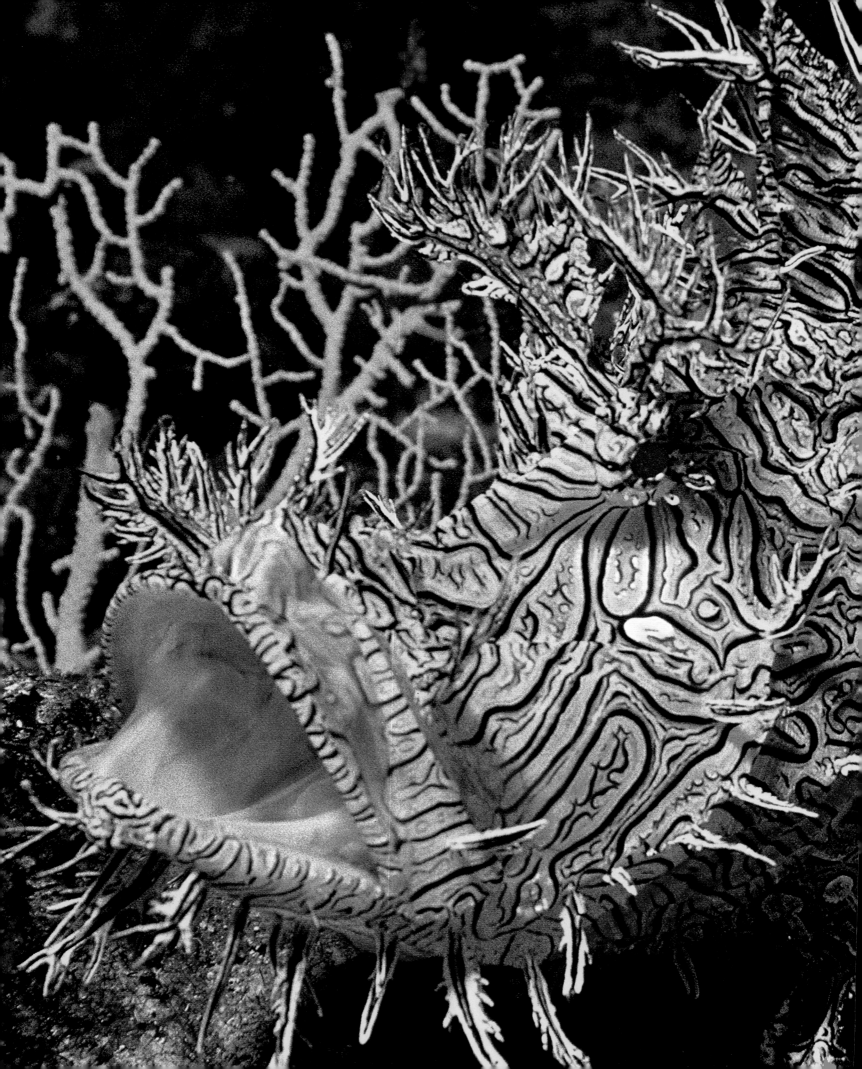

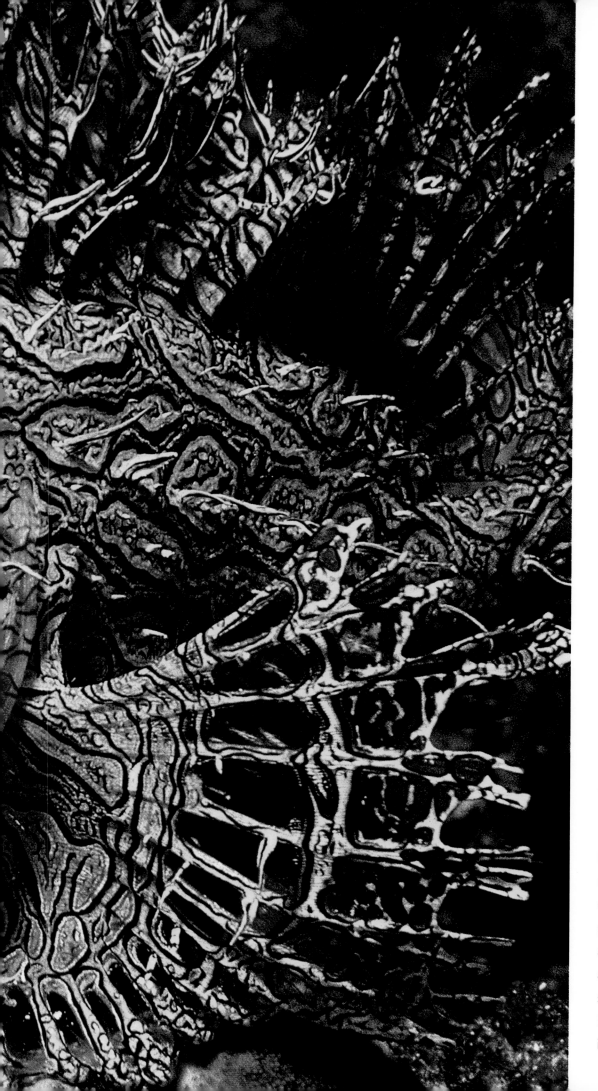

ON THE
SEA FLOOR

Most of the Earth's oceans are still
a mystery. In fact, it is estimated
that only about 5 percent of the
undersea world has been explored.
And species of marine life that
live along the sea floor or in the
greatest depths of the ocean are
seldom seen by humans. For
example, the carnivorous Merlet's
or Lacy Scorpionfish (left) is found
in the waters near Australia,
Japan, New Caledonia and Papua
New Guinea. Its remote habitat
combined with its complex pattern
of colouration and tentacles that
resemble algae – all of which
serve as camouflage – mean that
it is rarely sighted. In the extreme
depths of the sea, knowledge of
marine life is far from complete.
But humans are beginning to
learn more about this hidden
realm and some of the strange
creatures that live there – like
giant tube worms, which are found
as deep as 10,000 feet (3,050m),
around hydrothermal vents, where
sulphurous water, heated by the
Earth's molten interior, escapes
into the ocean.

SCULPTING PARADISE Fernando de Noronha, a volcanic island off the coast of
Brazil, is part of an archipelago whose contours have been smoothed by the waves.
Most of it is now protected as a National Marine Park.

BETWEEN LAND AND SEA

The ocean's waters are constantly eating away at the coastlines of the continents and the edges of islands. The incessant action of crashing waves, powerful currents and rising and falling tides, combined with the effects of wind and rain, sculpt the coast and help determine whether sandy shores, rocky cliffs or vast tidal flats will develop and how they will evolve. The effects of the sea on coastal topography can be relatively rapid, and not just in geologic time. This is especially true on beaches, which generally form in areas where the continental shelf is relatively wide and are made up of loose materials deposited along the coastline, such as quartz sand or fragments of bone and shell deposited by the sea. Because waves almost always strike the shore at oblique angles, beach sand and other fragments are continually being transported up or down the coast. In some cases, strong currents can wash away a beach in just a few years, though the same forces can also build up sand on other shores.

Where the continental shelf is narrow or is angled steeply downward and where rivers and streams do not carry large amounts of sand and silt toward the sea, rocky coastlines are more common. In this type of environment, which is often the site of complex tectonic landforms or previous glacial activity, waves usually hit the coast with more force. Harder rock, like granite, tends to resist the eroding effects of the ocean, resulting in the formation of cliffs, while softer rock, such as shale, will result in more gradual slopes. However, since there is often more than one type of rock in any given area, erosion will often be unequal, creating a coastline with jagged projections of land, sea caves and blocks of stone dislodged by the waves.

ISLANDS ARE ESPECIALLY VULNERABLE TO THE EROSIVE ACTION OF THE OCEAN SINCE THEY ARE COMPLETELY SURROUNDED BY ITS TIRELESS WATERS. SEVERE OCEAN STORMS OFTEN GIVE ADDED POWER TO THIS PROCESS, AND RISING SEA LEVELS ARE ONLY ACCELERATING THE DISAPPEARANCE OF ISLAND TERRAIN.

CLIFFS AND COASTLINES
Rocky coasts provide some of the starkest and most spectacular settings in nature. With waves pounding against the bottom of high cliffs, they highlight the seeming insignificance of humans in the face of the massiveness of the stone and the power of the sea that can wear it away. But if the hardest rock resists the

longest, in this battle of titans the water will win in the end. Water can carve a magnificent arch out of solid stone, like this one at Etretat on the coast of Normandy (France), and, as water continues its work, even an arch will eventually collapse, leaving only a column, known as a sea stack, as the last visible outpost of an eroding coastline.

ATOLLS AND LAGOONS
When a volcano in the ocean becomes extinct and begins to sink, its life is not necessarily over. In warm tropical waters, a coral reef can form around the volcano's perimeter, slowly creating an atoll. Growing upward and outward toward the open sea, the surface level of coral eventually becomes separated from the volcano by

water, resulting in a ring-shaped formation encircling a lagoon of usually about 100–200 feet (30–60m) in depth. If the volcanic base has not sunk too far or is not completely eroded, remnants may jut out in the centre of the lagoon, such as seen here at Bora Bora, a two-peak extinct-volcano island with a coral atoll that is part of the Society Islands in French Polynesia.

CRASHING THROUGH THE WAVES Sailing on the open water is one way in which humans continue to test themselves against the power of the seas – and escape to a world of elemental beauty.

MASTERING THE SEA

Humans have long been in awe of the power of the sea. But they have also been working for millennia to harvest its riches and harness its great potential. Fishing is one of the oldest employments in the world, and the ancient Egyptians ventured out into the Mediterranean and the Red Sea as early as the third millennium BCE. The Age of Exploration (1420s–1600s) and the mass migration of Europeans to the New World would, of course, have been impossible without fleets of sturdy ships and reliable navigational techniques.

Today, the sea provides the world's shipping with its most important highway. Approximately 60 percent of the oil consumed on the planet is transported by tankers, a method also used for liquefied natural gas. Overall, about 90 percent of all goods traded in the world are carried by sea (right, the scene in Kowloon Bay, Hong Kong). And if the aviation era means that the golden age of maritime passenger transportation is over, its spirit is being revived in a new generation of luxury ships, like the Queen Mary 2, which entered service in 2004. While transatlantic service is still being offered on some of these new ships, they are primarily designed for vacation cruises rather than as a means of getting from one place to another.

Finally, the ocean's tides, currents and waves – as well as temperature gradients – may one day provide electricity for homes and industry. Experiments are already underway to exploit these potential energy sources, which could help reduce the world's dependence on fossil fuels and nuclear power.

THE CONTINUING DOMINANCE OF OCEAN TRANSPORT HAS BEEN MADE POSSIBLE BY THE DEVELOPMENT OF CONTAINER SHIPS, BEGINNING IN THE 1960S, WHICH VASTLY INCREASED THE EFFICIENCY OF THE CARRYING OF NON-BULK CARGOES. INDEED, WITHOUT CONTAINERIZED SHIPPING, IT IS UNCERTAIN WHETHER SOME OF THE RECENT ECONOMIC BOOMS IN COUNTRIES LIKE CHINA WOULD HAVE HAPPENED.

FISHED ALMOST SINCE THE BEGINNING OF TIME, THE OCEANS ARE A MAJOR SOURCE OF THE WORLD'S FOOD SUPPLY. BUT OVEREXPLOITATION OF THIS VALUABLE RESOURCE AND THE WASTEFUL DESTRUCTION OF LARGE NUMBERS OF MARINE ANIMALS EACH YEAR HAVE PUT THE GLOBAL FISHERIES IN PERIL.

HARVESTING THE SEAS

Humans have searched for food in the Earth's waters for thousands of years, beginning in sheltered coastal areas and then venturing out further and further into the open sea. Today, marine fisheries exist in every ocean, with the most productive areas being in the northwest and southeast Pacific Ocean. Saltwater fish and other seafood, most of which is caught in coastal waters, account for the vast majority of fishery products consumed throughout the world. In fact, currently averaging between 80 and 90 million tons a year, the world's oceans and seas account for about 90 percent of the global catch, with the rest being made up from inland fishing. Though often part-time or seasonal, fishing activities employ almost 30 million people worldwide. And fish, crustaceans and molluscs contribute significantly to feeding the planet's ever-growing population. The annual average consumption is about 35 pounds (16kg) per person per year, though this figure varies considerably from region to region. In many developed countries the amount is about double the global average. The fruits of the sea represent a valuable part of many people's diet, providing micronutrients, minerals, essential fatty acids and protein. But not everything harvested from the world's seas ends up on the plates of humans. About a quarter is used in manufacturing fishmeal and oil – products made from Peruvian anchoveta and Atlantic menhaden among other species – and as direct feed for animals.

The global fisheries, however, are not an inexhaustible resource. It is estimated that three-quarters of fish stocks are currently exploited at maximum or excessive levels. In fact, overfishing – combined with the effects of ocean pollution – has caused a critical decline in the populations of many species. Commercially attractive large fish such as tuna, cod and swordfish have seen their numbers decline by as much as 90 percent over the last century. One result of this situation is that some fishermen are now targeting smaller species, which are lower on the food pyramid. This shift could have disastrous effects on the ecological balance of ocean environments.

BEYOND THE LIMIT (LEFT)
Some fishing practices are particularly destructive and indiscriminate, with wasteful, not to mention barbaric, consequences. Sharks are often hunted solely for their fins; the poor use of nets by industrial fleets (left) results in thousands of dolphins and porpoises being trapped and killed unintentionally each year. Though governments are working to regulate the world's fisheries in order to ensure a continued supply for future generations (and especially a livelihood for smaller scale traditional fishermen – right), the task is enormous.

FISH FARMING One response to overfishing and the depletion of the world's stocks is aquaculture, the cultivation of fish and other seafood, as well as seaweed, in special tanks, pools or enclosured sections of natural bodies of water – here (above), pens in Lake San Isabel on Luzon island, Philippines, are used to farm tilapia. Saltwater fish farming, a fairly recent

development, is known as mariculture; it is becoming increasingly popular, notably in China, and represents about 35 percent of the current aquaculture production. Mariculture includes the raising of aquatic plants; molluscs such as clams, oysters, scallops and mussels; crustaceans, including shrimp, crabs and lobsters; and fish such as Atlantic salmon and sea bass.

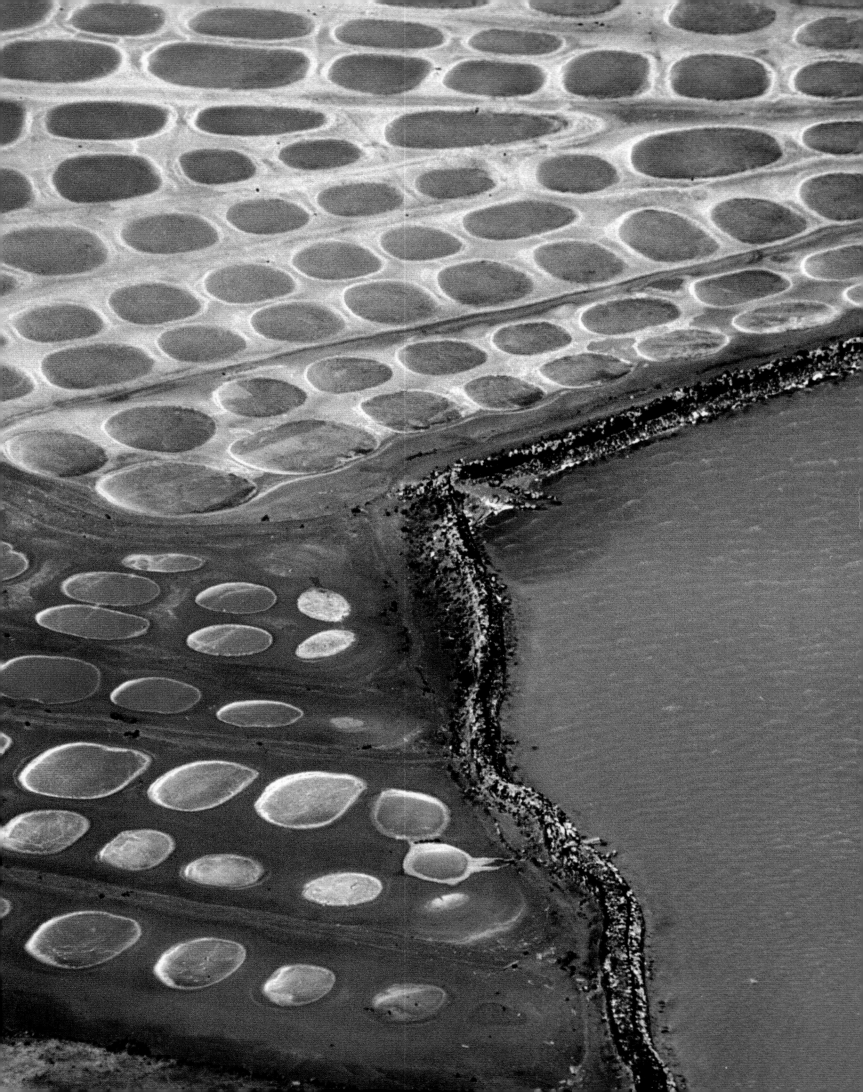

THOUGH THE SALTY TASTE OF SEAWATER COMES FROM A COMBINATION OF SUBSTANCES, THE COMPOUND SODIUM CHLORIDE, OR COMMON SALT, IS PRESENT IN THE LARGEST CONCENTRATIONS. FOR THOUSANDS OF YEARS, HUMANS HAVE TURNED TO THE SEAS AS A SOURCE FOR THIS MOST BASIC OF MINERALS.

SALT FROM THE SEA

Salt is an essential part of the human diet, and its importance was recognized early on in history. In the Roman army, soldiers were given a salt allowance, or *salarium*. This is the origin of the English term "salary". And there is more salt in the world's oceans than anywhere else – an estimated 4.5 million cubic miles (18.8 cu. km) of it. In fact, almost all salt used for human consumption used to come from the evaporation of seawater. Thus, in 1930 when Mahatma Gandhi launched a non-violent movement against the British authorities in India by leading thousands of people to the sea to make salt, in defiance of the government's salt tax, he was, from one point of view, simply inviting them to return to an activity that had been practised for millennia.

Over time, rock salt (halite) mining largely replaced the earliest methods, but in regions such as Guérande in Brittany (right), Trapani in Sicily and Lanzarote in the Canary Islands (left), where there is access to seas or salt lakes, the old ways continue and drying is still by solar evaporation. Such natural methods are even making a comeback as a gourmet, and perhaps healthier, alternative to industrially produced table salt.

The traditional process of producing salt begins by feeding water from a nearby sea or salt lake into shallow ponds where the evaporation process gets started and impurities are removed. During this first stage, vivid colours often develop in the ponds due to varying algal concentrations. These colours, which can be striking – and almost like abstract art when seen from a distance – are also dependent on the salinity of the water, which increases, of course, as the process advances. In ponds with low to medium salinity, greens predominate, while in high-salinity ponds the colour shifts to red. Bacteria can also contribute to the changing hues. Evaporation ponds often attract a wide variety of waterfowl, which feast on the algae and tiny brine shrimp present in the water. Once the salt concentration is high enough, the resulting brine is transferred to a series of crystallizing pans where the drying and purification steps are completed.

STILL WORTH ITS SALT (LEFT)
A stone-lined saltpan on the island of Lanzarote in the Canaries recalls the days when salt was widely used in the preservation of fish. Though its production on the island has diminished, salt is still extracted from the sea in the traditional manner.

DESALINATION

In regions where freshwater is scarce, desalination of seawater can provide an alternative source. Distillation, using a heating process to create water vapour, which is then condensed, is the most common technique for removing salt and impurities, though other methods, such as reverse osmosis, also exist. Desalination is expensive,

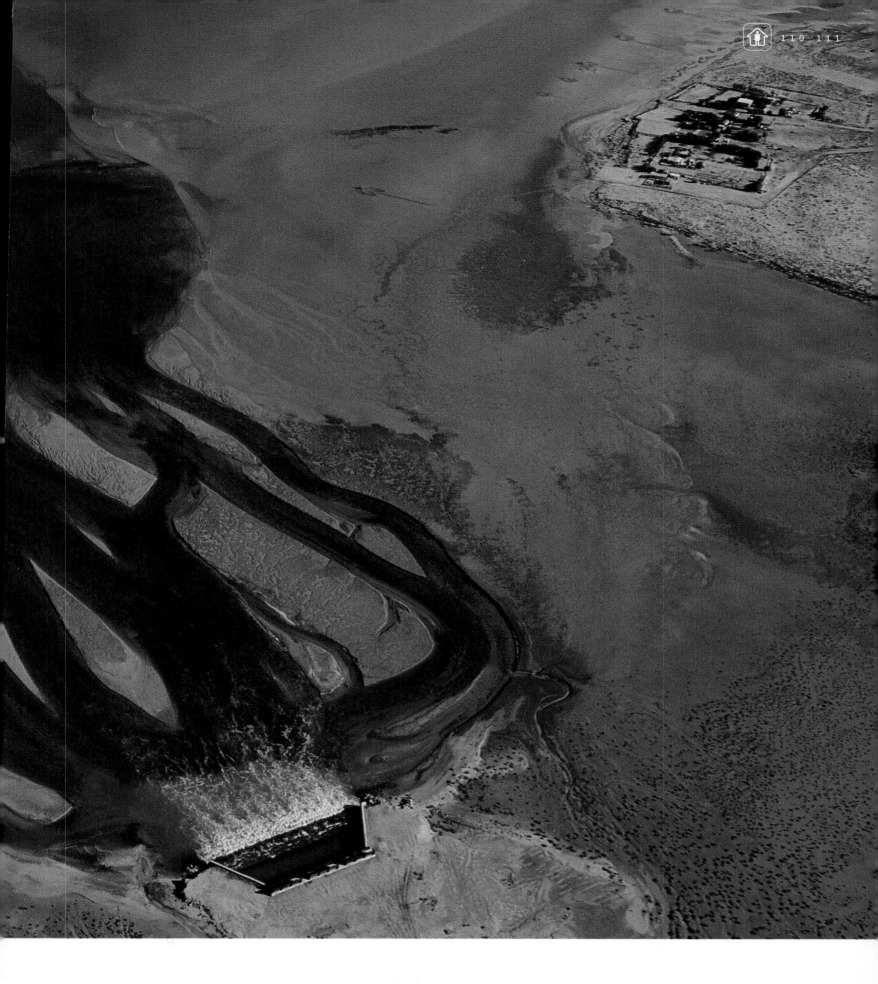

however, and oil-rich countries of the Middle East are among the few that can afford this form of water production on a large scale – though, with diminishing costs, its use is increasing. Thermal desalination plants – like thermal power stations – produce wastewater as a by-product, such as can be seen being discharged from this plant at Al-Doha in the Jahra region of Kuwait.

THE WORLD'S OCEANS ARE SICK. HUMANS HAVE BEEN POURING TONS OF WASTE INTO THE SEAS AS IF THERE WERE NO TOMORROW – BUT TOMORROW HAS ARRIVED. URGENT MEASURES NEED TO BE TAKEN TO PROTECT MARINE ECOSYSTEMS BEFORE THE DAMAGE BECOMES IRREVOCABLE.

THE AILING OCEANS

For far too long, people have been treating the Earth's oceans like one big dumping ground, and the result is a poisoned aquatic world, littered with debris that threatens virtually all species of marine life and the entire ecosystem of the planet's largest living space. Most of the pollution that ends up in the sea comes from land-based activities. Runoff from agriculture, industry and various domestic sources, laden with a host of harmful substances, enters rivers, streams and groundwater, eventually finding its way to the oceans. Over 300 million gallons (360 million US gal, 1,360l) of oil from industrial waste and automobiles, for example, enters the sea every year. Acids, scrap metals, waste from fish processing and coal ash also end up in the sea. And pesticides and fertilizers from farms add to the lethal mix. Dredging from the construction and maintenance of rivers, canals and harbours constitutes the bulk of the waste dumped into the sea. This enormous quantity of silt and sludge sometimes contains heavy metals, hydrocarbons and excessive levels of nutrients, including phosphorous and nitrogen. And even when dredged material is "clean", it can disrupt fish and crustacean spawning areas.

Oil spills from tankers are another major source of ocean pollution. Disasters like the ones caused in 1989 by the *Exxon Valdez* off the coast of Alaska and the *Prestige* near Spain in 2002 get extensive media coverage, but an estimated 600,000 barrels of oil are accidentally spilled into the sea every year. These events have dire consequences for the environment, often resulting in the deaths of thousands of birds and other marine animals and the long-term disruption of coastal ecosystems. A more common source of pollution from ships comes from the 10 billion tons of ballast water that is dumped back into the sea every year. This water often contains traces of oil and various chemicals and can also introduce non-native species – such as the zebra mussel or comb jellyfish – into a new environment, destabilizing the local ecology. Marine debris is also having tragic consequences. Everything from soda cans to plastic bags to derelict fishing nets cause the injury and death of many creatures, such as pelicans, sea turtles and dolphins, that ingest or get entangled in these discarded materials. The situation of the world's oceans is not hopeless. Stricter anti-pollution laws are being passed, and better methods for preventing and cleaning up oil spills have been developed in recent years – but more action is needed, and time is running out.

TERRIBLE BEAUTY
A thin layer of oil spreads out on the water. Experts can estimate the extent of an oil spill by its hue. Rainbow-coloured bands indicate a thickness of about 0.000012 inches (0.0003mm), which represents about 170 gallons (200 US gal) per square mile or 2.9 litres per hectare.

_Aether Aurium

THE SKIES

3

Where is the rain going to come from? The sky, of course. But will there be enough of it, and will it fall in the right places? The answer, I am sorry to say, is probably not. The Millennium Development Goals, agreed by the United Nations, aim to reduce by half the proportion of people who suffer from hunger. This means, the UN says, that the world must double its use of water for growing food between now and 2050. But how? Already many of the poorest parts of the world are suffering from an acute scarcity of water, and they use almost all of it to grow crops. The latest scientific projections suggest that, far from doubling, available water is likely to decline in the places that need it most. In the dry parts of the world, according to the Intergovernmental Panel on Climate Change, freshwater will decrease by between 10 and 30 percent by the middle of the century. Rainfall will decline and the glaciers and snowpack

_ by **GEORGE MONBIOT**

ENVIRONMENTAL JOURNALIST

which irrigate some of the most important food-producing regions will disappear. In Africa alone, the panel says, up to 250 million people could face increased water stress by 2020. What can be done about this? We must use every tool we have to prevent runaway climate change. We must find better ways of catching and saving water and irrigating crops. But we might also have to prepare for the worst: a world without enough rain, which means a world without enough food.

SUN

CLOUD
AND WATER VAPOUR

EVAPORATION
AND TRANSPIRATION

PRECIPITATION

EARTH

AN ENDLESS CYCLE The water cycle is the process by which water travels from the
Earth's surface to the atmosphere and then back to the ground again. It is constantly
in motion, with the same water going through the cycle over and over again.

THE WATER CYCLE

When it rains, the water that falls from the sky is part of the same water that has been on Earth for millions of years. The planet, a closed system, never gains or loses water, but circulates it in an endless cycle of evaporation and precipitation. Some of it moves through this dynamic process rather quickly – in a matter of weeks – and some of it can take thousands of years if it happens to be in the depths of the oceans or becomes trapped in a glacier.

Since the hydrologic cycle is, in fact, circular, it doesn't really have a beginning or an end, but let us join the cycle at the point when, through the warming power of the sun, water evaporates from the oceans and from the Earth's rivers, lakes and streams. Plants also add moisture to the air through a process known as transpiration. All this water vapour eventually condenses in the atmosphere, forming clouds. After about a week, precipitation, whether as rain, hail, sleet or snow, brings the water back to Earth. Most of it falls into the sea, while some adds to the polar icecaps. The rest replenishes the freshwater reserves of the continents.

The water that falls on land can follow many paths before it finds its way back to the atmosphere. Some of it will evaporate quickly, and plants also absorb precipitation before transpiring some of it into the air. Rain and snow feed the planet's rivers, streams and lakes, and much of this water will eventually make its way to the sea. A certain proportion of the precipitation will infiltrate the soil and rock and add to the groundwater, which sooner or later will find its way back to the surface or to the world's oceans, there to await its return to the atmosphere.

> "BETWEEN EARTH AND EARTH'S ATMOSPHERE, THE AMOUNT OF WATER REMAINS CONSTANT; THERE IS NEVER A DROP MORE, NEVER A DROP LESS. THIS IS A STORY OF CIRCULAR INFINITY, OF A PLANET BIRTHING ITSELF."
> *LINDA HOGAN*

TRANSPIRATION
ACCOUNTS FOR ABOUT
10 PERCENT OF THE WATER VAPOUR IN THE ATMOSPHERE.

A LARGE OAK TREE TRANSPIRES
OVER 30,000 GALLONS
(36,000 US GAL, 136,000L)
OF WATER A YEAR.

AN ACRE OF CORN TRANSPIRES
ABOUT 3,000 GALLONS
(3,600 US GAL, 13,600L)
IN A SINGLE DAY.

CLOUDS, THE ATMOSPHERE'S FRESHWATER RESERVOIRS, FORM IN A MULTITUDE OF SHAPES AND SIZES AND CAN TRANSFORM THE SKIES INTO SPECTACLES OF BREATHTAKING BEAUTY OR DARK AND THREATENING BACKDROPS THAT ANNOUNCE THE ARRIVAL OF STORMS, FLOODING AND DESTRUCTIVE WINDS.

CLOUD FORMATION

Clouds form when water vapour, purified by the process of evaporation and transpiration, rises with a parcel of air and begins to cool. If condensation occurs, the vapour can become visible as a cloud, which is actually a mass of water droplets, ice crystals or a combination of the two. Various conditions can cause a humid air mass to rise; for example, it may encounter the side of a hill or mountain, or it may be forced upward by an approaching front. Normally, cloud formation occurs at relatively high altitudes, sometimes up to 50,000 feet (15,000m), though low-lying fog can also be considered a type of cloud.

As the air mass encounters areas of lower pressure at higher altitudes, it begins to expand. This expansion cools the parcel through an adiabatic process (the enlarging of the mass of humid air requires energy, or work, which takes heat away without, in this case, there being any transfer of heat from the surrounding air). The cooling occurs in a direct relation to the cloud's rising, called the "lapse rate," which is about 5.4°F (9.7°C) for every 1,000-foot (305m) increase in altitude. When the temperature finally falls below the dew point, supersaturated air is formed. The moisture in the air can then begin to condense on microscopic particles, known as condensation nuclei, forming droplets less than a thousandth of an inch in diameter. Once the concentration of these droplets reaches a few thousand per cubic inch, the cloud becomes visible. Usually some of the droplets will become so cold that they will turn into ice crystals, especially in the higher parts of the cloud.

As long as the air current continues to move upward and precipitation does not occur, a cloud will remain in the sky, majestically aloof. However, the reverse of the cloud-forming process can also occur. If a cloud descends, and thus moves into zones of higher pressure, it will begin to decrease in size and its temperature will rise. Warm air can hold more moisture than cool air, and the water droplets and ice crystals will be forced back into the form of water vapour. Thus, when clouds sink they start to evaporate before any moisture ever returns to the Earth's surface. In fact, only about one-fifth of the water vapour held in clouds actually ever falls as rain or snow.

THE COLOURS OF CLOUDS (LEFT)
Because water droplets are highly reflective, they give clouds a white appearance when viewed from above or with the sun at your back. However, when seen from below, clouds appear grey or black as they reflect and scatter sunlight away from the viewer. Sometimes, this same light-scattering property can give clouds a wide range of spectacular hues, especially at sunrise or sunset.

ALL SHAPES AND SIZES
Clouds come in a great variety of types and can be classified in several different ways. The four basic forms are cirrus, nimbus, cumulus and stratus. Cirrus are thin, high-level clouds, usually associated with fair weather. Nimbus clouds, which generally form at an altitude of about 7,000 to 15,000 feet (2,100 to 4,600m),

are usually dark and bring steady precipitation. Cumulus are white, billowy clouds that reveal the action of thermal uplift in what look like vertically piled-up heaps of fluffy cotton balls. Stratus clouds are relatively featureless formations that can blanket an entire sky, bringing dull, grey weather. The bases of these clouds are often only a few hundred feet above the ground.

WHEN THE FOG CREEPS IN
Fog is actually just a cloud close to the Earth's surface and forms in essentially the same way as other cloud types. The main difference is that the cooling of the moist air mass necessary to trigger condensation does not come from an increase in altitude, except in the case of unslope fog, which forms when a parcel of

air moves up the side of a hill or mountain. Radiation fog forms over land, usually on clear nights, when cooler ground chills low-lying humid air. Building from the bottom, this type of fog can reach heights of 500 feet (150m) or more. When relatively warm, moist air passes slowly over a colder wet surface, such as a bay (like San Francisco, above) or ocean, advection fog can form

"RAIN IS GRACE; RAIN IS THE SKY CONDESCENDING TO THE EARTH; WITHOUT RAIN, THERE WOULD BE NO LIFE." *JOHN UPDIKE*

THE OPENING UP OF THE HEAVENS

When a cloud just can't hold any more moisture, the rain comes tumbling down. But something has to get the process started, since the water droplets in clouds are usually too small and light to fall to earth. Their size can grow in two main ways. When a cloud's tiny water particles, which are all moving about at different speeds, collide, they coalesce and begin to form larger and larger drops. When these become too heavy to remain suspended in the air, they fall as rain. In colder clouds where both water droplets and ice crystals are present, the droplets can adhere to the crystals causing them to grow. While these crystals could fall as snow, very often they melt into rain before reaching the ground.

One way meteorologists classify rain is by its hourly rate of fall. Light rain is generally defined as less than one inch (2.5cm) per hour, moderate rain between 1 and 3 inches (2.5 and 7.5cm), and heavy rain over 3 inches (7.5cm) an hour. In a downpour the figure can go much higher. In 2007, for example, during a tropical storm, Wake County, Virginia reported 2.2 inches (5.6cm) of rain in just 20 minutes. Different parts of the world, of course, receive different amounts of rain. Mount Waialeale (whose name means "Rippling Water"), on the Hawaiian island of Kauai, is one of the wettest places on Earth, with an average annual rainfall of 460 inches (1,168cm); in 1982, it received a record 666 inches (1,692cm) of rain.

Not all precipitation falls as rain. Besides snow, water can return to Earth in other forms like sleet and hail. Sleet occurs when rain falls in areas where the surface temperature is low, causing it to freeze on the way down. Hail is formed during thunderstorms through the effect of updrafts, which carry ice particles that have collided with water droplets to the higher parts of the clouds. There they refreeze, forming larger and larger pellets, depending on how many times they are carried back up. When the hailstone finally gets too heavy, it falls to the ground. In the American Midwest, hailstones as large as 6 inches (15cm) in diameter have been observed.

RAINDROPS (RIGHT)
Contrary to popular belief, raindrops are not shaped like tears. Small raindrops are almost perfectly spherical, and larger ones tend to be flattened on the bottom and dome-shaped on top. If they get too big, they break up into several small drops as they fall.

ACID RAIN (LEFT)
Sulphur dioxide and nitrous dioxide emitted by automobiles, factories and power stations can combine with water vapour in clouds to form acids. When precipitation occurs, the acid rains contaminate lakes and rivers, damage plants and animal life and eat away at buildings.

A SPECTRUM OF LIGHT

Created by the refraction and reflection of sunlight as it passes through water droplets, rainbows, which may appear in the mist of a waterfall or the spray of a wave as well as in the sky near the end of a rainstorm, display all the colours of the visible spectrum. Each colour is bent at a slightly different angle. Red, with the longest wavelength, appears on top, followed by orange, yellow, green, blue, indigo and violet. Sometimes there is a fainter secondary bow, resulting from a double reflection of light within the water droplets, in which the order of the colours is reversed. A dramatic effect can occur if part of the sky is still dark with clouds when the rainbow shines out in the sunlight. Rainbows seem magical, in part because they don't actually exist in a particular place. Where you see them and what they look like depends not only on weather conditions, but also on the observer's location and the position of the sun. And they never last very long.

MORNING DEW
When relatively warm, moist air passes over cool objects near the ground, it can condense and form droplets of dew. This often occurs on clear nights, when exposed surfaces lose their heat more easily to the sky. Dew is also more likely to form on objects that are not in direct contact with the warming effect of the ground, like grass or spider

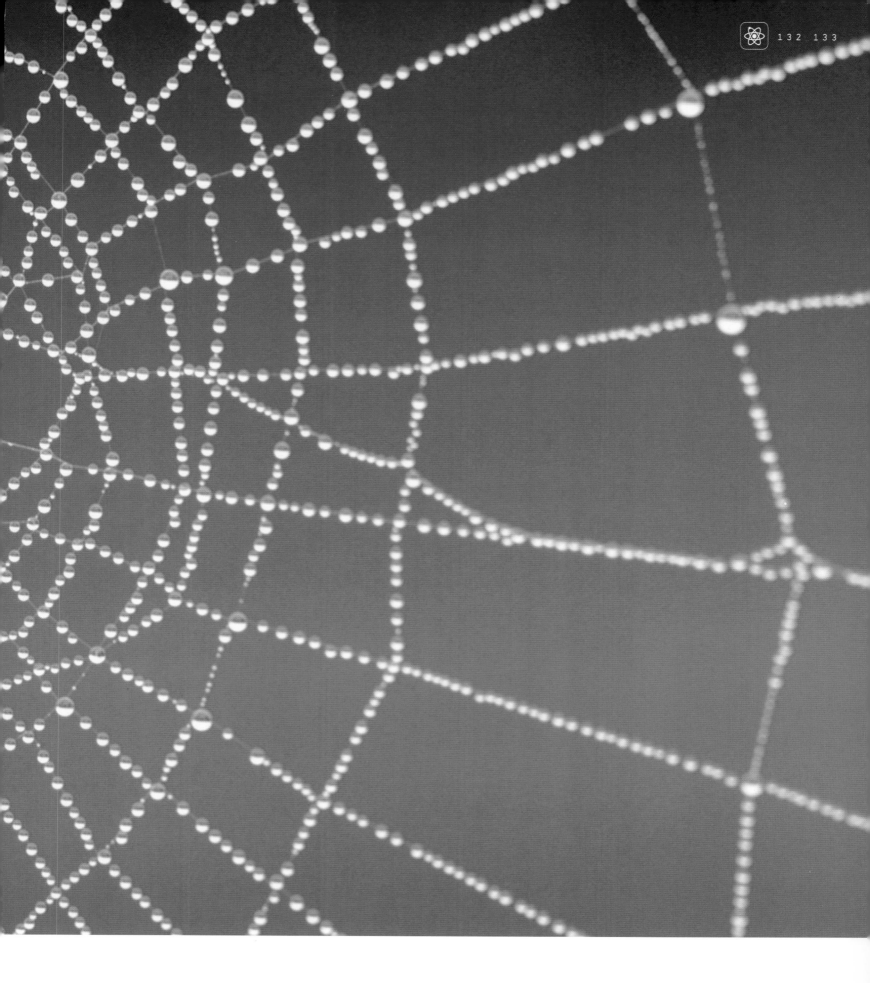

webs. In regions where rainfall is scarce, dew harvesting can supply an additional source of freshwater, which can be collected from the droplets that form on metal or tile surfaces, such as roofs, or captured on large nets placed on coastal mountainsides, a technique used in Chile and Peru.

WEATHER SATELLITES In use since the 1960s, weather satellites reveal large-scale cloud formations, giving meteorologists a comprehensive view of the Earth's weather situation that enables more accurate forecasts – here, a hurricane heads for the USA.

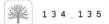
THE WORLD'S WEATHER

Since it's always changing – and because it can bring welcome relief or terrible disaster – humans have always been concerned with the weather. And in the form of rain, hail, sleet or snow, water is what many people focus on when they think about weather. Of course, wind, temperature variations and air pressure all have an effect on atmospheric conditions, but it is the oceans that really drive the global weather machine. Their capacity to absorb, store and release heat as well as moisture helps determine just how the weather will change, whether it will vary relatively frequently, like in most temperate zones, or less so, as in many tropical regions.

Weather occurs in the lower region of the Earth's atmosphere, known as the troposphere, though sometimes activity at higher altitudes can affect meteorological conditions. Air masses – bodies of air that have a relatively uniform temperature and moisture level throughout – are crucial in determining an area's weather. The point of contact between two of these parcels of air is called a front, and it is here that precipitation normally occurs. For example, if a cold front moves into an area where there is a warmer air mass, it will tend to push that body of air upward, thus favouring cloud formation and perhaps rain, thunderstorms or snow.

Since the weather has such a major effect on the lives of so many people, humans have always striven to predict what it will do next. If certain scientific methods began to be introduced in the nineteenth century, it was the advent of radar in the 1940s that revolutionized weather forecasting. An extremely effective tool to detect precipitation, it was enhanced with the development in the 1980s of Doppler radar, which can detect the movement of areas of precipitation.

CLOUD SEEDING INVOLVES INTRODUCING SUBSTANCES THAT CAN SERVE AS CONDENSATION NUCLEI INTO CLOUDS TO INDUCE PRECIPITATION. CHINA HAS PRACTISED THIS TECHNIQUE BY FIRING SILVER IODIDE ROCKETS INTO THE SKIES ABOVE ARID REGIONS IN AN ATTEMPT TO BOOST AGRICULTURAL PRODUCTION.

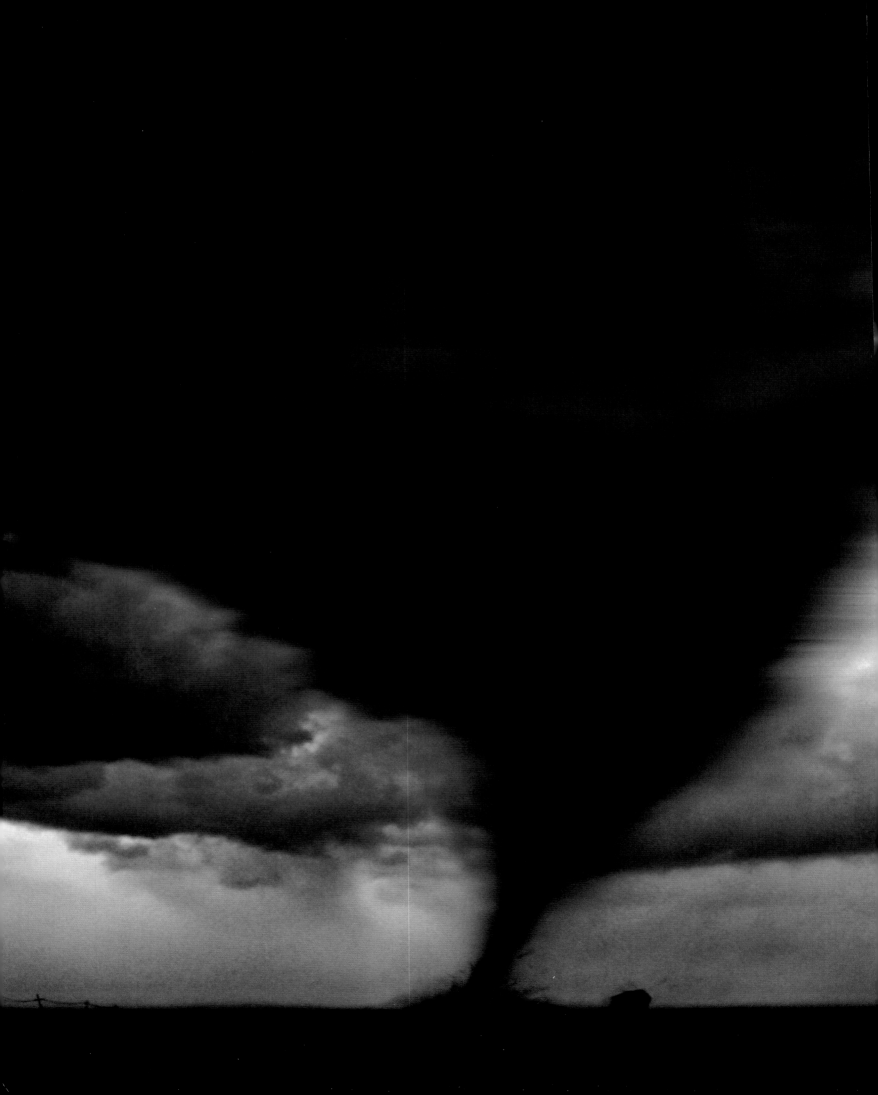

THE WRATH OF THE SKIES

When it combines with strong winds, water can be frightening and destructive. Thunderstorms are the most common form of severe weather, with a global average of up to 40,000 every year. These violent, though short-lived, phenomena are produced by cumulonimbus clouds. Causing heavy rain, often accompanied by hail, the clouds also become charged with electricity. This is released in the form of lightning (see previous pages), which produces shock waves that are heard as thunder. Tornadoes sometimes occur in association with thunderstorms. The whirling air in a tornado's vortices (left) can reach up to 300 miles (480km) an hour and cause spectacular damage, though areas just outside its path may remain untouched.

THE GREENHOUSE EFFECT IS A NATURAL PHENOMENON THAT HELPS KEEP THE EARTH WARM ENOUGH TO SUSTAIN LIFE, BUT HUMAN ACTIVITY HAS LED TO THE EMISSION OF VAST QUANTITIES OF GASES THAT RAISE GLOBAL TEMPERATURES AND MAY HAVE DISASTROUS CONSEQUENCES FOR HUMANITY.

THE GREENHOUSE

A layer of gases in the atmosphere absorbs heat reflected from the Earth, and about 30 percent is radiated back toward Earth again. This blanket of atmosphere, acting somewhat like the glass in a greenhouse, keeps the average global temperature at about 60°F (15.6°C). Of the naturally occurring greenhouse gases (GHGs), water vapour makes up the largest percentage. Without these gases, the planet would be a subfreezing 0°F (-17.8°C), too cold for most life.

Human activity, especially since the Industrial Revolution, has added to the greenhouse effect. By increasing the amount of gases that help trap heat in the atmosphere, the world's population is contributing to a rise in the planet's average temperature. The burning of fossil fuels (coal, oil and natural gas) is the main reason for this warming, though other factors play a part as well. Fluorocarbons, nitrous oxide and methane are also produced by human activity and add to the effect. The digestive process of ruminant livestock such as cows and sheep, for example, contributes about 30 percent of human-related methane emissions. Overall, global emissions of GHGs due to human activity have increased by 70 percent between 1970 and the present.

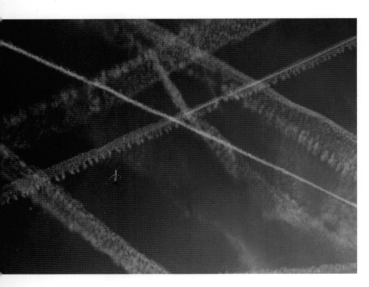

While there is still some disagreement about the exact rate of global warming and its precise effect on the planet, a general consensus has arisen over the last few years, notably about its link with human activity. In 2007, the Intergovernmental Panel on Climate Change reported cautiously, but firmly, that, "Most of the observed increase in globally-averaged temperatures since the mid-20th century is very likely due to the observed increase in anthropogenic GHG concentrations." Global warming is already having an effect on the planet's weather and will likely lead to more extreme conditions, including droughts, flooding, heatwaves, high winds and severe storms. The greenhouse effect is also contributing to the melting of the ice caps and rising sea levels, a trend which could have catastrophic effects for much of the Earth's population.

THE EARTH'S BLANKET (RIGHT)
As the sun's rays warm the Earth's surface, much of the heat energy is radiated back towards space as infrared light. Naturally occurring gases in the atmosphere – including water vapour, carbon dioxide, methane and nitrous oxide – absorb some of this infrared energy and keep it from escaping into outer space.

CONTRAILS (LEFT)
Contrails are white streaks across the sky created by aeroplanes flying at high altitudes. Composed of condensed water vapour, they last from a few minutes to several hours, often transforming themselves into wispy cirrus clouds. These vapour trails add to aviation's already significant contribution to global warming.

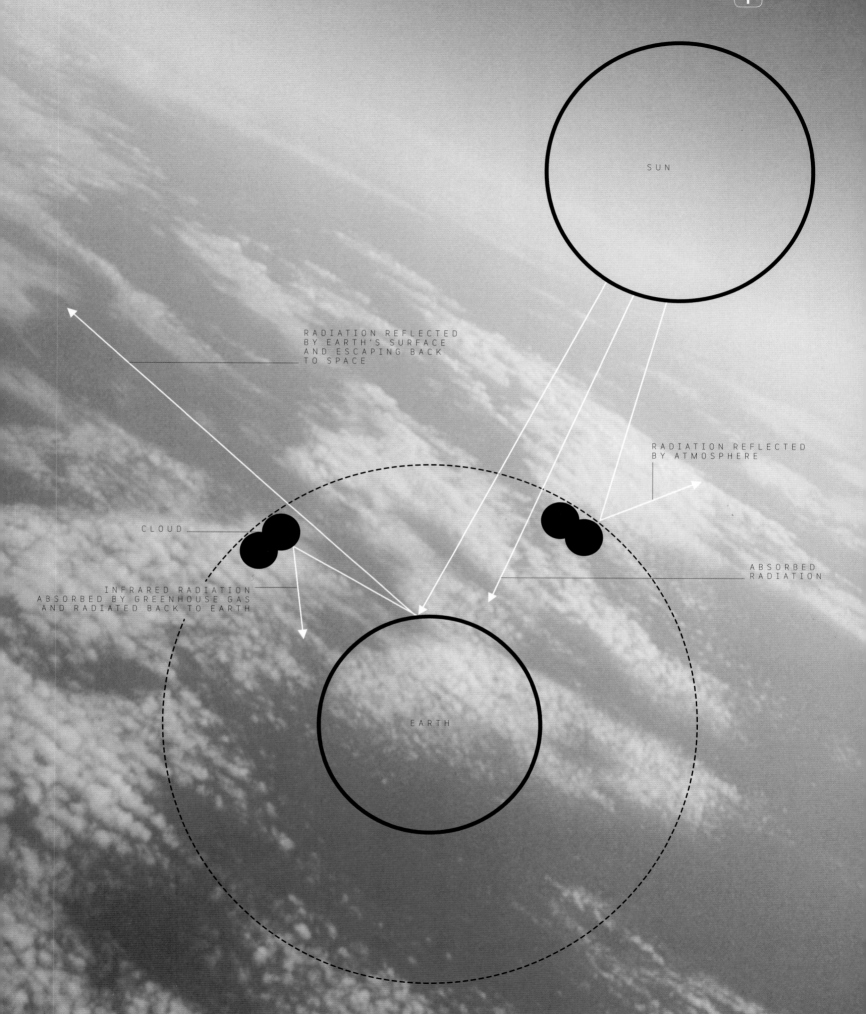

SUN

RADIATION REFLECTED
BY EARTH'S SURFACE
AND ESCAPING BACK
TO SPACE

RADIATION REFLECTED
BY ATMOSPHERE

CLOUD

ABSORBED
RADIATION

INFRARED RADIATION
ABSORBED BY GREENHOUSE GAS
AND RADIATED BACK TO EARTH

EARTH

STEAM LOCOMOTIVE Still in use in many places today, such as China (above), the steam locomotive was invented in 1803 by Richard Trevithick in England. It did not prove practical initially because the cast iron rails then in use were too brittle for its weight.

THE POWER OF STEAM

Steam is simply water vapour created by heat. It exists naturally, for example as part of phenomena like hot springs and geysers, but humans have also produced steam artificially for well over two centuries for a wide variety of purposes. The steam engine, perfected in 1765 by the Scot James Watt, harnessed this powerful energy in a useable and efficient manner and gave a substantial boost to the Industrial Revolution, which up until that time had relied primarily on watermills and windmills. By the early nineteenth century this invention was being adapted for use on boats and railways, with the American Robert Fulton designing the first passenger steamboat in 1807 and the Englishman George Stephenson proving the commercial viability of the steam locomotive in 1829 with his *Rocket*. In the United States, with its great distances, steam transportation was embraced enthusiastically, and by 1840 the country already had more miles of railroad track than all of Europe. Steam, in fact, radically transformed both transportation and industrial production throughout the nineteenth and twentieth centuries.

If the reign of steam was shaken in the early twentieth century by the advent of the internal combustion engine (diesel locomotives, for example, began to come into use as early as the 1920s), even after World War II steam continued to find other applications, such as the heating of factories and homes. Modern domestic uses also include cooking (with pressure cookers or specially designed food steamers), cleaning (with machines for washing curtains, rugs and carpets) and, of course, ironing. In industry, steam is employed today in the production of metals such as steel, aluminium and copper, in the manufacture of chemicals, and in petroleum refineries. And one of the most important roles of steam in contemporary society is in the production of electricity.

IN THERMAL POWER STATIONS THROUGHOUT THE WORLD, STEAM – WHETHER IT IS PRODUCED USING HEAT FROM A NUCLEAR REACTOR OR FROM THE BURNING OF COAL, OIL OR NATURAL GAS – HELPS CREATE ELECTRICITY. ITS THERMAL ENERGY IS CONVERTED INTO MECHANICAL ENERGY TO DRIVE TURBINES, WHICH GENERATE THE ELECTRIC POWER.

THE FRO

ZEN REALM

4

It is incredible to think that glacial ice covers one-tenth of the planet. I once spent three weeks in the Himalayas, climbing up the Bagini glacier and then on to the summit of Changabang. Low down, I marvelled at tiny, deep pools of water that stared out from the blue-green skin of ice and negotiated the network of melt-water streams that gushed into a smooth, dark throat in the ice, disappearing into a subterranean tangle of veins. Higher, the temperature plummeted and I witnessed the solid phase of water in a multitude of forms: powder snow, nevé, sastrugi, rime, icicles, hail, ice fields, and verglass. Finally, I sat poised below a fragile, knife-edge ridge winding up toward the summit, perched on a snow- and ice-bound mountain chain that runs almost 1,500 miles (2,500km) from Afghanistan in the west to Bhutan in the east. At that moment, I understood perfectly why this cold, lofty world is believed by many to be the home of the

_ by **ANDY CAVE**

MOUNTAINEER

gods. Below, through swirling cloud I could make out the glacier flowing down into the Ganges, a river that provides water for 500 million people. Almost 70 percent of the Earth's freshwater is held in icecaps and glaciers, but the frozen landscape is being altered significantly by global warming. At current rates, scientists claim that this could result in the disappearance of all Himalayan glaciers within 50 years. Glacial lakes are steadily growing and if they burst the result would be catastrophic flooding of the heavily populated regions below, followed eventually by a permanent drought. Antarctica has as much ice as the Atlantic Ocean has water and here ice is melting at an even faster rate.

Andy Cave

SPARKLING IN THE SUN Icicles slowly melt in the spring next to Namtso, or Nam Co, in east central Tibet, one of the country's three holy lakes. At an altitude of 15,478 feet (4,718m) above sea level, it is the highest saltwater lake in the world.

ICE IN NATURE

Water turns to ice when the temperature drops below 32°F (0°C). This natural phenomenon occurs all over the planet, though primarily at latitudes well above and below the equator or at high altitudes. The chill of the night also favours the formation of ice. One of the more unusual properties of water is that, unlike most substances, it actually expands when it changes from a liquid to a solid. Frozen water molecules form hexagonal crystals, which are about 8 percent less dense than water. This is why ice floats.

Ice occurs in nature in a wide variety of forms. It falls from the sky as snow, hail and sleet, and sometimes liquid precipitation solidifies upon impact with the ground to create freezing rain. Water vapour in humid air can deposit ice on cold surfaces, in the form of frost, without passing through a liquid phase. When the temperatures remain low long enough, rivers and lakes can also freeze over. Seawater can even turn to ice, but because of its salinity, among other factors, the temperature has to be below 32°F (0°C). When it is so cold that water in the ground remains frozen all year long, it creates permafrost making soil as hard as rock.

Near the poles, vast ice sheets of continental proportions cover Antarctica and much of Greenland. These formations have built up primarily through the accumulation of snowfall over thousands of years, the same process that has created ice caps, smaller masses of perennial frozen water. Mountain glaciers form in high altitude valleys and move down toward the flatlands below. When large pieces of ice shelves or glaciers break off and move out to sea, most frequently in spring and summer, they form icebergs.

MUCH LIKE STONE FORMATIONS IN CAVES, WHEN ICICLES BUILD UP THROUGH CONTINUED DRIPPING AND REFREEZING OF WATER, THEY CAN FORM MASSIVE STALACTITES, LIKE THESE PHOTOGRAPHED IN THE ALPS. THEY SOMETIMES BECOME LONG ENOUGH TO TOUCH THE GROUND AND CREATE COLUMNS OF FROZEN WATER.

"THE FIRST FALL OF SNOW IS NOT ONLY AN EVENT, IT IS A MAGICAL EVENT. YOU GO TO BED IN ONE KIND OF A WORLD AND WAKE UP IN ANOTHER QUITE DIFFERENT, AND IF THIS IS NOT ENCHANTMENT THEN WHERE IS IT TO BE FOUND?" *J.B. PRIESTLEY*

BLANKETS OF SNOW

When it's cold, water vapour in clouds condenses directly into ice crystals. Once enough of these crystals have combined, they can become heavy enough to fall as snow. The size of a snowflake depends on how many crystals it contains. Though most are less than half an inch (1.3cm) across, in relatively warm conditions, close to freezing, they can become much

larger. Individual ice crystals are clear, but snow appears white because it reflects so much light and because what little light is absorbed, is absorbed relatively uniformly across the spectrum.

Though technically it's never too cold to snow, heavier snowfalls tend to occur when there is relatively warm air (typically between 15°F and 32°F, -9.4°C and 0°C) near the ground, since it can hold more water vapour than colder air. And if fresh snow can create a winter wonderland, blizzards can dump huge quantities of snow in a relatively short period and cause severe damage. The so-called "lake effect" can greatly increase snowfall. Rochester, New York, not far from Lake Ontario, gets close to 100 inches (250cm) of snow a year, but regions further north can get even more.

Though people often think that there is a fixed water equivalent for a given amount of snow, the ratio can actually vary quite a bit. Depending on its density, 10 inches (25cm) of snow, for example, corresponds to anywhere from about a half to one inch (1.3–2.5cm) of rainfall. Snow, like all precipitation, forms part of the Earth's water cycle. However, since it falls in solid form, there is a slowing effect in its path through the cycle. If it melts soon after reaching the ground, then it will quickly evaporate or join a stream or river on its way to the ocean. However, in cold regions, the pause in the cycle may last until the beginning of the spring or summer. Finally, snow that falls on ice shelves or glaciers eventually becomes packed into hard ice and can remain in a solid state for centuries.

STILL LIFE (RIGHT)
Snow and frost cover a tree in the mountainous Black Forest region, along the eastern bank of the Rhine in southwestern Germany.

THE GREAT NORTH (LEFT)
In the extreme northeast of Norway, not far from the border with Russia, sand dunes on the Varanger peninsula are blanketed with snow.

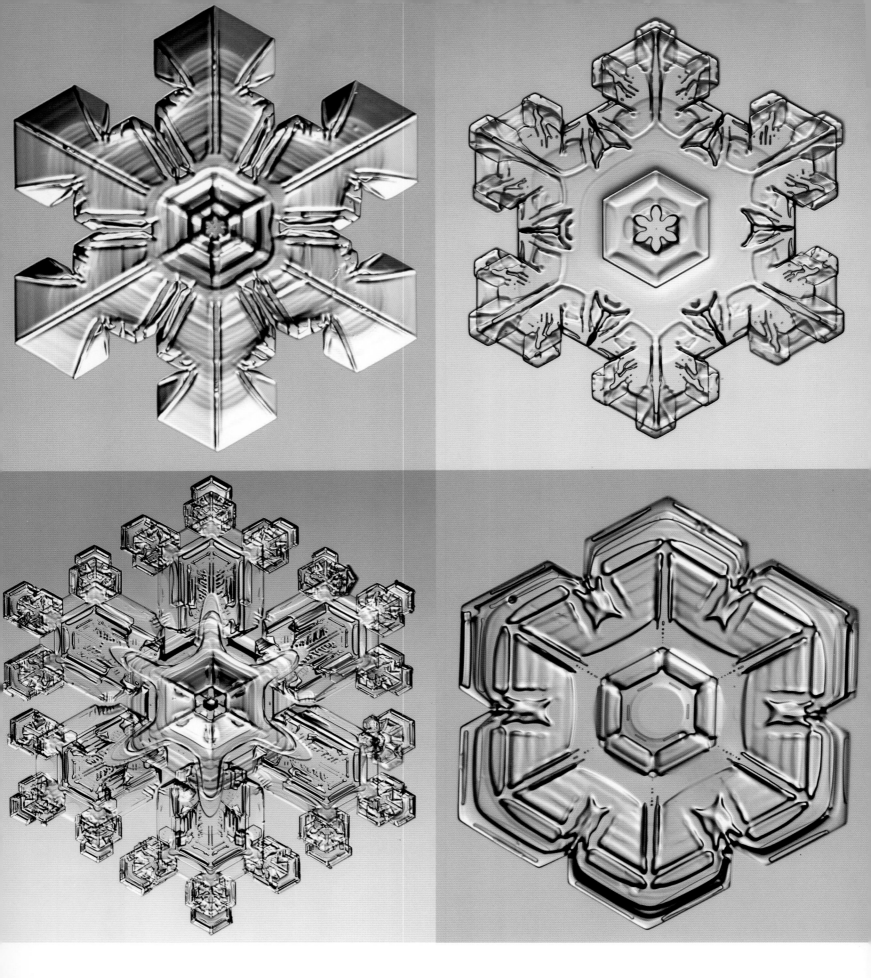

SNOW FLAKES
When water freezes, it forms six-sided crystals of a near infinite variety and stunningly intricate geometrical beauty. It is the agglomeration of these crystals that creates the snowflakes that fall from the sky, though the term is also used for the individual crystals. The symmetry of snowflakes is due to the hexagonal structure taken on by water

when it turns to ice, but in fact most crystals are not perfectly regular. Not all snowflakes are flat and star-like with dendrites – or delicate arms – either. Triangular configurations are also possible, and some crystals form in the shape of needles, hollow columns or solid prisms. How the crystal develops is determined by factors including temperature and humidity levels.

A RIVER OF ICE A glacier in Denali National Park in south central Alaska flows
around a protruding section of rock in its slow movement down the mountainside.

GLACIERS

A glacier is any large mass of perennial ice that forms from the build-up of snow or other precipitation and shows evidence of deformation and flowing due to inclines in terrain or the pressure of its own weight. Glaciers exist on every continent except Australia. They cover about 10 percent of the planet's surface and contain about 70 percent of the global freshwater supply. Though this general definition includes virtually all terrestrial accumulations of frozen water that do not completely melt each summer, a distinction based on size is often made between ice sheets, ice caps and mountain glaciers.

Ice sheets are the largest formations. The term is generally restricted to the two vast expanses of ice on Antarctica and Greenland, which blanket the land with a thick layer of frozen water. Ice caps are smaller, usually high-altitude masses formed when ice fills neighbouring mountain valleys and joins together. (The specific term "polar ice cap", however, refers to large high-latitude regions covered with ice, similar to, but smaller than, ice sheets. These include ice masses that form over water, which are not technically glaciers.) The smallest bodies of ice, which form in individual valleys, are referred to as mountain glaciers, or very often simply as glaciers.

NEW SNOW FALLS ON A GLACIER EACH YEAR. THE SNOW THAT SURVIVES PAST THE MELT SEASON BECOMES MORE ROUNDED AND COMPACT AND IS REFERRED TO AS FIRN. AS ADDITIONAL LAYERS ARE ADDED, THE FIRN IS COMPRESSED AND FUSES WITH THE GLACIER'S LOWER MASS OF ICE.

Mountain glaciers are like big conveyor belts that slowly move ice and rock down the slopes. In the upper section is the accumulation zone, where at least some of the previous winter's snowfall persists throughout the summer and eventually becomes part of the glacial ice. Further down is the ablation zone where all of the previous winter's snow melts before the onset of the next winter. The boundary between these two areas, the equilibrium line, can shift up or down the mountain from year to year, depending on snowfall and temperature patterns.

ICE IN MOTION
Glaciers appear to be solid as a rock, but as their thickness builds, gravity and pressure cause them to flow as if they were liquid. Upper layers of ice may move faster than lower ones, or basal sliding may occur, in which the glacier advances as a whole over the underlying terrain. Here, like a frozen river in slow motion, the Perito Moreno

Glacier in Los Glaciares National Park, Argentina, creeps downslope. About 20 miles (32km) long and 250 feet (75m) thick, it is fed by the Southern Patagonian Ice Field in the Andes. Pieces of this glacier often break off at its terminus and fall into Lake Argentino, in a process known as calving.

THE FORMATION OF THE EARTH'S FJORDS REQUIRED THOUSANDS OF YEARS OF GLACIAL ACTIVITY FOLLOWED BY THE RISING OF SEA LEVELS AFTER THE LAST ICE AGE. TODAY THEY REMAIN AS THE STARK AND SPECTACULAR RESULT OF THESE ENORMOUS NATURAL FORCES.

ICE-CARVED VALLEYS

Narrow inlets, usually along rocky coasts, fjords are carved out by glaciers moving down mountainsides. These formations were created during recent glacial cycles when large ice sheets covered much of the globe and when sea levels were lower. Often extending many miles inland, fjords can be very deep. Since Norway has such a high concentration of fjords, it's not surprising that the Norwegian word for this geological feature is used in English. It is related etymologically to the word "ford". And if a ford is shallow, while a fjord is deep, they are both places where people would cross a body of water to avoid making a long detour.

Several characteristics of fjords have helped geologists determine that only glacial activity could have created these dramatic formations. Their depth reveals the powerful scraping action of thick, heavy glaciers that were able to erode away so much soil and rock. The U-shape of fjord valleys, with steep walls, is also typical of glacial erosion, unlike valleys cut by rivers, which tend to be V-shaped. Also, fjords are normally deeper in their middle and upper reaches, and more shallow closer to the open sea. This unusual feature results from the fact that glaciers

have more erosive power higher up the valley, before they lose much of their weight through melting, and because when they reach the end of their path, they deposit the soil and rock fragments they had been carrying along with them in what is known as a terminal moraine.

At the end of the last ice age, when glaciers all over the world melted, sea levels rose and these valleys carved out by glaciers were filled with water, creating today's fjords. Because of the fjord valleys' steep walls, the water may be hundreds of feet deep very close to the shore. And streams often pour over the sheer drops creating towering waterfalls. The sea, of course, has not totally immersed the valleys, which continue up into the mountains, where there may be a small glacier. A river also sometimes forms in the upper valley, in the same place as the one that had cut the original valley before the glacier altered its shape.

FJORDS ON THE PACIFIC (LEFT)
The Tracy Arm Fjord, over 30 miles (48km) in length, lies in southeast Alaska. The site's pristine beauty is now part of a protected wilderness area managed by the US Forest Service.

RUGGED SPLENDOUR (RIGHT)
Icebergs float in an isolated fjord in front of snow-capped mountains near Scoresbysund, on the east coast of Greenland, above the Arctic Circle.

GLACIAL LAKES

Water from melting glaciers often forms lakes in the mountains. Since each year's melting process is relatively slow, especially at high altitudes, these lakes often do not reach their highest levels until early summer. Protected from most pollution, water in glacial lakes is exceptionally pure. However, in today's world, even these isolated bodies of water can be tainted by airborne toxins brought from distant sources. Nestled in the Canadian Rockies, the awe-inspiring Moraine Lake (left) lies at an elevation of over 6,000 feet (1,800m). A dramatic backdrop is provided by the Valley of the Ten Peaks, which includes some of the highest mountains in the Canadian Rockies. The blue-green colour of the lake comes from the reflection of the sky and the refraction of small particles of rock, known as glacial flour, suspended in the water, which are created by the grinding force of the glacier and carried to the lake by streams.

ICE PACKS As summer approaches, this ice pack, or frozen patch of sea, photographed
from far above the Earth's surface, begins to break up into small icebergs.

POLAR REGIONS

At the extreme ends of the Earth's axis lie the North and South Poles – the coldest regions on the planet. Each has a winter season of six months of almost complete darkness, when life seems literally frozen in time. In Antarctica, the mean annual temperature is around -58°F (-50°C), but in the winter it can get much colder. In July 1983, a Russian research station recorded a temperature of -129°F (-89.4°C). There is, of course, no landmass directly at the North Pole itself, but temperatures in the Arctic also drop to extreme lows in winter months. These two regions account for about 90 percent of the world's frozen freshwater. The rest is spread out in mountain glaciers and ice caps throughout the world.

The planet's two largest glacial formations, the ice sheets of Antarctica and Greenland, are located in the polar regions. Unlike glaciers in mountainous settings, whose flow is determined largely by the sloping terrain, ice sheets slowly spread out from the centre, forced by the sheer weight of their enormous mass, which builds every year due to winter snowfalls. Almost the entire continent of Antarctica is covered with ice, whose average thickness is over a mile. This accumulation of frozen water gives Antarctica a higher mean altitude than any other continent, a situation that only reinforces the cold climate. And the ice actually extends out beyond the land onto the open sea. The Greenland Ice Sheet does not cover the entire territory of this huge island, but it is nonetheless enormous. Smaller ice caps in places like northern Canada and Russia, cover much of the landmasses in the region around the North Pole, and much of the Arctic Ocean itself is frozen over for most of the year.

THE HARSH ENVIRONMENT OF THE POLES HAS FASCINATED AND CHALLENGED HUMANS FOR CENTURIES. IT WASN'T UNTIL THE TWENTIETH CENTURY, HOWEVER, THAT SERIOUS EXPLORATION OF THESE EXTREME REGIONS REALLY BEGAN. IN 1909 THE AMERICAN ROBERT E. PEARY REACHED THE NORTH POLE, THOUGH HIS CLAIM HAS BEEN DISPUTED, AND IN 1911 THE NORWEGIAN ROALD AMUNDSEN WAS THE FIRST PERSON TO MAKE IT TO THE SOUTH POLE.

ANTARCTICA

ITS NAME MEANS
"OPPOSITE TO THE ARCTIC".

IT IS THE ONLY CONTINENT WITH
**NO INDIGENOUS
POPULATION.**

IT HAS SO
LITTLE PRECIPITATION
THAT IT IS TECHNICALLY
A DESERT.

ITS ICE SHEET IS OVER
TWO MILES THICK
IN SOME PLACES.

ICE SHELVES
When glaciers, under the pressure of their own incredible weight, extend out over the sea, they create ice shelves. These formations, which remain attached to the land-based glaciers that sculpt them, are found only in Antarctica, Greenland and Canada. Resembling thick platforms, they often have sheer walls, left by huge masses of ice which

break off and fall into the sea. Though they float on the ocean's saltwater, they themselves are made up of freshwater. The Fimbul Ice Shelf, which is about 120 miles (190km) long and 60 miles (95km) wide, is located on Princess Martha Coast in Antarctica. Because of its exposed position over the deep ocean, it tends to lose more of its ice to melting than some larger ice shelves.

SEA ICE

Sea ice forms directly on the water, primarily in the Arctic Ocean and around the coast of Antarctica. On average, almost 10 million square miles (26 million sq. km) of ocean are covered by sea ice, which is more than twice the size of Canada. Since saltwater freezes at about 28.8°F (-1.8°C), rather than 32°F (0°C), sea ice – sometimes referred to as pack ice – requires colder temperatures before it begins to crystallize. Throughout the winter months, sea ice thickens and spreads out. During Antarctic winters, this phenomenon more than doubles the effective size of the continent, keeping the more inland regions isolated from the moderating effect of the ocean. In the summer, much of the sea ice melts, though in some areas part of it remains frozen all year round. If it is not too thick, specially designed icebreaking ships can plough a path through sea ice in order to reach remote areas.

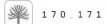

ORIGINALLY FORMED ON LAND FROM THE BUILD UP OF SNOW, ICEBERGS, WHICH SLOWLY MELT AS THEY FLOAT IN THE OCEAN, RETURN THEIR LOCKED-UP WATER TO THE SEA, AND TO THE ETERNAL WATER CYCLE.

MOUNTAINS IN THE SEA

When the ends of expanding glaciers break off and fall into the sea, they form icebergs. This process is called calving, and, in effect, glaciers give birth to icebergs. The majority of icebergs are created in the spring and summer, when warmer weather and melting facilitate both the flow and break up of glaciers. In the Arctic region about 10,000 are produced each year from the West Greenland Glacier alone. The size and shape of icebergs vary greatly. When they break off ice shelves in Antarctica, they tend to be fairly regular, with flat tops. Icebergs formed in Greenland, in contrast, are often more jagged or twisted, reflecting the rougher path they take to the sea. Arctic icebergs don't usually get as large as those in the south, about 150 feet (45m) tall and 500 feet (150m) long being a typical size. Still, veritable ice islands of up to a mile or more in length have occasionally been spotted. The Antarctic is the home of not only the largest number of icebergs, but of the real monster-size specimens. Lengths of five miles (8km) are not unusual among these tabular-shaped platforms.

With a density less than that of water, icebergs, as is well known, float. But most of an iceberg, about seven-eighths, remains below the surface. This hidden aspect is one reason that icebergs can be so dangerous to ships. They are particularly treacherous when they drift far from their normal latitudes. Most Arctic icebergs, for example, stay above 55° N. And while a whole series of mistakes – not to mention just plain bad luck – contributed to the infamous sinking of the *Titanic* in 1912, the fact that such a large iceberg had drifted as far south as 42° N, approximately the same latitude as Rome, clearly made its appearance less expected. Today, detection methods have been vastly improved, and icebergs are occasionally towed away from oil platforms or shipping lanes in order to reduce the risk of accidents. The fact that icebergs are made of frozen freshwater has led some people to see them as a potential source of drinking water, but such use has not yet proved economically feasible.

FLOATING MAJESTY (LEFT)
The uneven melting of an iceberg near Petermann Island, off the west side of the Antarctic peninsula, creates a stunning frozen arch in the sea

TWISTED ICE (OVERLEAF)
Icebergs calved directly into the sea by continental glaciers, rather than by more regularly formed ice shelves, can have strangely contorted and eerie shapes.

ECOSYSTEMS IN POLAR REGIONS ARE NOT AS DIVERSE AS THOSE IN TEMPERATE AND TROPICAL CLIMATES. STILL, MANY SPECIES OF ANIMALS HAVE BEEN ABLE TO ADAPT TO THESE HOSTILE ENVIRONMENTS, AND A FEW MAKE THE ARCTIC OR ANTARCTIC THEIR YEAR-ROUND HOME.

LIFE AT THE EXTREMES

In general, the further north or south on the globe one goes, the smaller the number of species one finds. No amphibians or reptiles, for example, live near the North or South Poles, and some animals spend the summer in these areas, but migrate to warmer latitudes in the winter. Still, some species have been able to survive in these extreme regions. One feature common to many polar animals is their relatively stocky appearance and layers of fat, both of which help retain body heat.

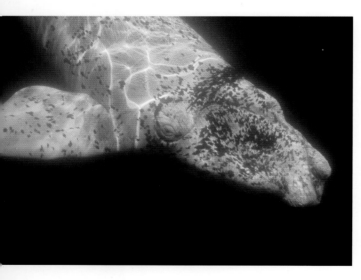

About 20 species of mammals and 100 species of birds live in the Arctic region, though not all of these venture into the extreme northern latitudes. Perhaps the most emblematic inhabitant of the North is the polar bear, the largest land-based predator on the planet, whose white colour provides the perfect camouflage for surviving in the Arctic. Polar bears rely on the sea for much of their food. This close link with the water is reflected in their scientific name, *ursus maritimus*. The caribou, Arctic wolf and Arctic fox also spend at least part of their lives on the ice of the North. And marine mammals like seals and walruses thrive in Arctic waters. The beluga whale and the narwhal are peculiar to the region, and other whale species also frequent the Arctic.

Antarctica is more isolated from the rest of the world than the Arctic, a fact which has had an influence on its fauna. There are no native land mammals on the continent, though several species of birds do make their homes in the region during at least some months of the year. There are almost 20 different varieties of penguin. The largest of these, the Emperor penguin, breeds exclusively on the continent and nearby islands. The distinctive colouring of penguins is actually a form of protective camouflage. When they are in the water, their black backs do not stand out to predators looking down from above, and their white fronts tend to blend in with the sky for any enemy that might be lurking below.

THE DEEP SOUTH (LEFT)
Off the coast of Patagonia, Argentina, a southern right whale swims below the surface. As their name indicates, these baleen whales tend to stay in the lower latitudes of the globe.

COMING UP FOR AIR (RIGHT)
A humpback whale breaches in the cold waters of the Arctic. Humpbacks head to polar feeding grounds in the summer and migrate to tropical regions in the winter months to breed.

FAUNA ADAPTATION Polar animals present a striking example of adaptation, which has resulted in some of the planet's most distinctive species. A colony of Emperor penguins gathers here on an island off the coast of Antarctica, isolated from predators. The largest species of penguin, they can weigh up to 90 pounds (40kg). Besides whales, walruses are

the largest mammals of the polar regions, and males of the species can reach over 2,700 pounds (1,230kg). Walruses, who live in the Arctic seas, use their tusks to dig clams along the shoreline. Another inhabitant of the North, polar bears are good swimmers, a skill they rely on in hunting seals. Here a polar bear stretches himself on the ice near Churchill in Canada.

A SURGE IN MOUNTAIN WATER The Tsomoriri Lake in Ladakh, India, is fed
by melting snow and glaciers in the Himalayas. Global warming is leading to lakes
like this one receiving more water than they can handle, threatening serious flooding
in the spring and summer seasons.

A WARMING WORLD

The Earth's average temperature is increasing, and the effect of this trend is perhaps most dramatic in the world's frozen realms. In fact, as the Intergovernmental Panel on Climate Change noted in 2007, the planet is experiencing widespread melting of snow and ice, which is leading, among other things, to a rise in sea levels. The report stated that mountain glaciers and snow cover have declined in both hemispheres, and that the melting of the permafrost is creating increased ground instability. The great ice sheets are also diminishing, as are the many ice caps throughout the polar regions. The disappearance of the world's naturally formed ice masses may be the most immediately threatening consequence of global warming.

In many mountainous regions, local residents have observed the ongoing retreat of glaciers and the lessening of snowfalls for several years. Ski resorts, for example, are no longer certain they will be able to remain open for a full season. But that will probably be the least of the planet's problems. Greater runoff from mountains will increase the number and size of glacial lakes, upsetting ecosystems and leading to increased flooding. In the long run, the opposite effect – no less problematic – is likely to occur. As the frozen reservoirs of the mountains disappear, eventually there will be less and less water to feed lakes and rivers, the result then being droughts rather than floods. It is estimated that at current rates, most of the glaciers in the Himalayas will disappear within 50 years. One of the greatest worries is the impact that these changes will have on agriculture, where harvests could be devastated by too much – and then too little – water.

THE MULAJOKULL GLACIER IN CENTRAL ICELAND IS FED BY THE HOFSJOKULL ICE CAP (BELOW). NUMEROUS PONDS AND RIVULETS OFFER EVIDENCE OF AN INCREASED WATER FLOW FROM THE MELTING GLACIER.

BLUE LAKES AND RIVULETS

For several years now, a growing number of blue lakes and rivulets have been appearing in the Arctic snowcaps at higher and higher latitudes. Naturally occurring phenomena that form in the ice itself, their increasing presence is a consequence of global warming, and they actually have an accelerating effect on the melting of polar ice. Since the waters in blue lakes absorb more of the sun's heat than highly reflective ice or snow, they themselves cause further melting, which runs off in a growing number of rivulets. This blue lake (left) formed on a glacier in Alaska. Below, water from a glacier in Grise Fiord, Canada, situated over 700 miles (1,130km) above the Arctic Circle, heads toward the sea.

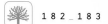

THE FABLED NORTHWEST PASSAGE, LONG SOUGHT BY EARLY EXPLORERS, IS NOW BECOMING A REALITY. SO MUCH OF THE NORTHERN POLAR ICE CAP HAS MELTED THAT IN THE SUMMER OF 2007 THIS NATURAL WATERWAY FROM THE ATLANTIC TO THE PACIFIC BECAME NAVIGABLE FOR THE FIRST TIME. IN JUST TWO DECADES THE NORTHWEST PASSAGE MAY BE USABLE ALL YEAR ROUND, BUT THE PRICE THAT THE PLANET IS PAYING FOR THE REALIZATION OF THIS OLD DREAM COULD BE VERY HEAVY INDEED.

MELTING ICE

The whole planet seems to be melting: snow cover, glaciers, permafrost, sea ice, ice shelves and the great ice sheets of Antarctica and Greenland. Since most of the world's frozen water is found in the polar regions, it's also there that the greatest melting is occurring. And the situation is being exacerbated in the Arctic by the fact that average temperatures appear to be rising faster there than in the rest of the world. The effect on the Arctic Ocean ice cap is already dramatic. It has lost over 20 percent of its size in the last 30 years, and the reduction seems to be picking up speed. And while melting in Antarctica may be proceeding at a somewhat slower pace, the impact there is nonetheless enormous. Huge tabular icebergs, often several miles in length, are dropping off the ice shelves surrounding Antarctica at increasing rates.

One of the most disturbing aspects of global glacial melting, in fact, is that – as in the case of blue lakes – it is a self-accelerating process. As ice melts, it eventually exposes either land or open ocean. Since both of these are darker and less reflective than ice or snow, they absorb more of the sun's energy, result-ing in greater warming and even faster melting. Rising sea levels are one of many potentially catastrophic consequences of all this added water. And since the planet and everything on it keep getting warmer, the seawater itself is expanding, taking up more room and adding further to the rise, though some scientists think that this effect will be partially offset by increased snowfall over Antarctica. The exact extent of future increases in sea level is difficult to predict, but the result could be the total submersion of some low-lying islands and coastal areas and significantly increased flooding in others. And the ocean's absorption of additional heat, as more and more of it is uncovered by melting sea ice and ice shelves, is affecting currents. Changes in the global movements of the ocean's water may have devastating effects on marine ecosystems as well as world weather patterns, which have already been affected by the warming of the atmosphere.

CRASHING INTO THE SEA (LEFT)
Huge blocks of ice break off a glacier and fall into the sea, adding more volume to the world's oceans.

MELTING AWAY (RIGHT)
Sea ice that forms in the Gulf of St Lawrence, Canada, usually reaches its maximum extent in early March, before beginning to melt and break up. In recent years it has been less thick and has tended to melt earlier.

SEA WORLD

Sea levels rose by about 200 feet (60m) following the end of the last ice age and stabilized around 2,000 to 3,000 years ago. The twentieth century, however, witnessed the onset of renewed sea level rising, with an increase between 1900 and 2000 of about 7 inches (18cm). And the seas are rising more quickly today than a hundred years ago. It is estimated that levels could rise by as much as 17 inches (43cm) by the end of this century. As global warming continues, melting glaciers will calve larger and more numerous icebergs, and major ice shelves will disappear. Among the animals the most vulnerable to these changes are polar bears and penguins who are having greater and greater difficulty finding food. Polar bears, generally strong swimmers, have been drowning since they have been forced to go farther and farther from the dwindling ice masses in search of prey. And the melting of sea ice in Antarctica is linked to a reduction in the amount of krill, a staple of the Emperor penguins' diet. If this gives them a shorter journey to the water in search of nourishment, the overall impact of the reduced food supply is a decrease in the population.

_GLOSSARY

Abyss The deepest part of the ocean.

Ablation (1) The process by which ice and snow waste away as a result of melting and/or evaporation. (2) The erosive processes by which a glacier is reduced.

Accretion The slow addition to land by deposition of water-borne sediment. An increase in land along the shores of a body of water, as by alluvial deposit.

Accumulation zone The area where the bulk of the snow contributing to an avalanche was originally deposited.

Acid rain Rain that has an abnormally low pH, due to contact with atmospheric pollutants such as sulphuric oxides.

Adiabatic process A change involving no gain or loss of heat.

Advection The process by which solutes are transported by the bulk of flowing fluid such as groundwater.

Algae Single- or multi-celled organisms, such as duckweed, that are commonly found in surface water.

Aquaculture The science and business of cultivating marine or freshwater fish or shellfish, such as oysters, clams, salmon and trout, under controlled conditions for commercial purposes.

Aquifer An underground geological formation from which groundwater can be extracted (using a well); its surface is the water table.

Archipelago A large group of islands.

Atmosphere The gaseous layer covering the Earth. The atmosphere is one of the four components, together with the lithosphere, hydrosphere and biosphere, that comprise the Earth's ecosystem.

Atoll Ring-shaped coral reef and small island that encloses a lagoon and is surrounded by open sea.

Basal sliding The process by which a glacier undergoes thawing at its base, producing a film of water along which the glacier then flows.

Basin A geographic area drained by a single major river; consists of a drainage system comprised of streams and often natural or manmade lakes.

Cenote A limestone sinkhole with groundwater.

Climate change *See* Global warming.

Cloud seeding A weather modification technique involving the injection of a substance into a cloud for the purpose of influencing the cloud's subsequent development.

Continental shelf The submerged shelf of land that slopes gradually from the exposed edge of a continent for a variable distance to the point where the steeper descent to the ocean bottom begins.

Contrail A visible trail of streaks of condensed water vapour or ice crystals sometimes forming in the wake of an aircraft.

Coral reef A ridge of limestone, composed chiefly of coral, coral sands and solid limestone resulting from organic secretion of calcium carbonate.

Coriolis effect The effect that the Earth's rotation has on moving bodies.

Delta Tract of land at or near the mouth of a river, resulting from the accumulation of sediment supplied by the river.

Dendritic A drainage pattern in which tributaries branch irregularly in all directions from and at almost any angle to a larger stream.

Depletion The water consumed within a service area or no longer available as a source of supply; that part of a withdrawal that has been evaporated, transpired, incorporated into crops or products, consumed by man or livestock or otherwise removed.

Deposit Something dropped or left behind by moving water.

Desalination The process of removing salt from seawater or brackish water.

Ebb tide The period of tide between high water and the succeeding low water.

Ecosystem An interacting network of groups of organisms together with their non-living or physical environment.

Effluent Water that flows from a sewage treatment plant after it has been treated.

Equilibrium line The point in a glacier where overall gain in volume equals overall loss, so that the net volume remains stable. The equilibrium line marks the border between the zone of accumulation and the zone of ablation.

Estuary A place where fresh and salt water mix, such as a bay, salt marsh or where a river enters an ocean.

Evapotranspiration The process by which water is discharged to the atmosphere as a result of evaporation from the soil and surface-water bodies and transpiration by plants.

Firn Old snow on the top of glaciers that has become granular and compact through temperature changes, forming the transition stage to glacial ice.

Fishery The aquatic region in which a certain species of fish lives.

Fjord A long, narrow, deep inlet of the sea between steep slopes.

Flood plain A strip of relatively flat and normally dry land alongside a stream, river or lake that is covered by water during a flood.

Fluoridation The addition of fluoride to a water supply to reduce tooth decay.

Freezing rain Rain which freezes upon contact with the ground.

Freshwater Water that contains less than 1,000 parts per million (1,000 milligrams per litre) of dissolved solids; generally, water with more than 500 parts per million (500 milligrams per litre) of dissolved solids is undesirable for drinking and many industrial uses.

Fumarole A hole or orifice in a volcanic region and usually in lava, from which issue gases and vapours at high temperature.

Geothermal activity Relating to the Earth's internal heat; commonly applied to springs or vents discharging hot water or steam.

Geyser A geothermal feature of the Earth where there is an opening in the surface that contains superheated water that periodically erupts in a shower of water and steam.

Glacial lake A lake that derives its water, or much of its water, from the melting of glacial ice; also a lake that occupies a basin produced by glacial erosion.

Glacier A large mass of ice, formed on land by the compaction and recrystallization of snow.

Global warming The progressive gradual rise of the Earth's surface temperature thought to be caused by the greenhouse effect and responsible for changes in global climate patterns.

Greenhouse effect The phenomenon whereby the Earth's atmosphere traps solar radiation, caused by the presence in the atmosphere of gases such as carbon dioxide, water vapour and methane that allow incoming sunlight to pass through but absorb heat radiated back from the Earth's surface.

Groundwater Subsurface water, such as in wells, springs and aquifers.

Gypsum A soft, white mineral composed of hydrous sulfate of lime.

Halite A colourless or white mineral found in dried lakebeds in arid climates, mined or gathered for use as table salt.

Hot spring A spring that brings subterranean hot water to the surface.

Hydrological cycle The natural recycling process powered by the sun that causes water to evaporate into the atmosphere, condense and return to earth as precipitation.

Hydropower Power produced by falling water; the utilization of the energy available in falling water for the generation of electricity.

Hydrosphere The part of the Earth composed of water including clouds, oceans, seas, ice caps, glaciers, lakes, rivers, underground water supplies and atmospheric water vapour.

Hydrothermal Having to do with hot water, especially having to do with the action of hot water in producing minerals and springs or in dissolving, shifting and otherwise changing the distribution of minerals in the earth's crust.

Ice pack A large area of floating pieces of ice driven relatively closely together.

Ice sheet A very large ice cap, also called a continental glacier, such as the one covering Antarctica.

Ice shelf Thick glacial ice that extends from glaciers on land and floats on the sea.

Ice cap An extensive dome-shaped or plate-like perennial cover of ice and snow that spreads out from a centre and covers a large area.

Lagoon A shallow stretch of seawater near or communicating with the sea and partly or completely separated from it by a low, narrow, elongated strip of land.

Mangrove Tropical evergreen trees and shrubs that have stilt like roots and stems and often form dense thickets along tidal shores.

Mariculture The cultivation of marine organisms for use as a food resource. *See also* aquaculture.

Membrane filtration A procedure used to filter

and purify water, usually without the addition of chemicals.

Monsoon A wind from specific directions that brings heavy rainfall to geographical locations in seasonal patterns.

Mudpot An acidic hot spring with a limited supply of water and large amount of hydrogen sulfide gas.

Oxbow lake A bow-shaped lake formed in an abandoned meander of a river.

Permafrost Any frozen soil, subsoil, surficial deposit or bedrock in Arctic or subarctic regions where below-freezing temperatures have prevailed continuously for two or more years.

Plankton Free-floating, mostly microscopic aquatic plants.

Quartz sand The most common rock-forming mineral.

Rainbow A circular bow or arc exhibiting, in concentric bands of light, the several colours of the spectrum and formed opposite the sun by the refraction and reflection of the sun's rays in drops of rain.

Rapids A part of a stream where the current is moving with a greater swiftness than usual and where the water surface is broken by obstructions, but without a sufficient break in slope to form a waterfall.

Recycled water Water that is used more than one time before it passes back into the natural hydrologic system.

Renewable water Water continuously renewed by the hydrological cycle.

Reservoir A pond, lake or basin, either natural or artificial, for the storage, regulation and control of water.

Rivulet A small stream or brook; a streamlet.

Runoff The part of precipitation water that runs off the land into streams or other surfacewater.

Safe water Water that does not contain harmful bacteria, toxic materials or chemicals and is considered safe for drinking.

Salt lake A landlocked body of water that has become salty through evaporation.

Sea ice Solidified water that forms when ocean or sea water freezes.

Sedimentation Settling of solid particles in a liquid system due to gravity.

Shallow A term applied to a shallow place or area in a body of water; a shoal.

Sinkhole A depression in the Earth's surface caused by dissolving of underlying limestone, salt or gypsum.

Solvent A substance that dissolves other substances, thus forming a solution. Water dissolves more substances than any other and is known as the "universal solvent".

Stalactite An icicle-shaped mineral deposit, usually calcite or aragonite, hanging from the roof of a cavern, formed from the dripping of mineral-rich water.

Stalagmite A conical mineral deposit, usually calcite or aragonite, built up on the floor of a cavern, formed from the dripping of mineral-rich water.

Steam vent *See* Fumarole.

Tailing Crushed rock that remains after processing ore to remove the valuable minerals.

Thermohaline circulation Global-scale overturning of the ocean driven by density differences arising from temperature and salinity effects. One of the best known examples of thermohaline circulation is the Gulf Stream.

Tidal range The difference between a tide's high and low water levels.

Tide The rhythmic, alternate rise and fall of the surface (or water level) of the ocean and connected bodies of water, occurring twice a day over most of the earth, resulting from the gravitational attraction of the moon.

Transpiration The process by which water passes through living organisms, primarily plants, into the atmosphere.

Tsunami A huge sea wave caused by a great disturbance under an ocean, such as a strong earthquake or volcanic eruption.

Updraft A warm column of air that rises and condenses within a cloud.

Wastewater Water that has been used in homes, industries and businesses that is not for reuse unless it is treated.

Water cycle *See* Hydrological cycle.

Water rights Legal rights to the use of water.

Water tower A standpipe or elevated tank used as a reservoir or for maintaining equal pressure in a water system.

_FURTHER READING

Most of the books listed are reference works but there are several popular non-fiction works and classic works of literature included in the list because they offer informative and compelling reading for anyone fascinated by the planet's seas.

Annin, Peter. *The Great Lakes Water Wars*. Island Press: Washington, D.C., 2006.

Aquado, Edward and James Burt. *Understanding Weather and Climate*. Prentice Hall: New Jersey, 2006.

Ball, Philip. *H$_2$O: A Biography of Water*. London: Phoenix, 2000.

Barlow, Maude. *The Global Water Crisis and the Coming Battle for the Right to Water*. New Press: New York, 2008.

Byatt, Andrew; Fothergill, Alastair; and Holmes, Martha. *The Blue Planet: A Natural History of the Oceans*. BBC Books: London, 2001.

Carson, Rachel. *The Sea Around Us*. Oxford University Press: Oxford, 1991.

Conrad, Joseph. *The Children of the Sea: A Tale of the Forecastle [The Nigger of the "Narcissus": A Tale of the Sea]*. Dodd, Mead & Co: New York, 1897.

Copeland, Sebastian. *Antarctica: The Global Warning*. Earth Aware Editions: San Rafael, California, 2007.

De Villiers, Marq. *Water: The Fate of Our Most Precious Resource*. Mariner: New York, 2001.

Earle, Sylvia A. *Fathoming the Ocean: The Discover and Exploration of the Deep Ocean*. The Belknap Press: Harvard, Massachusetts, 2005.

Ellis, Richard. *The Empty Ocean*. Island Press: Washington, D.C., 2003.

Emsley, John. *Nature's Building Blocks: An A–Z Guide to the Elements*. Oxford University Press: Oxford 2003.

Fennimore Cooper, James. *The Sea Lions*. (3 vols.) Bentley's Standard Novels: New York, 1849.

Field, John. G., Gotthilf Hempel and Colin. P. Summerhayes. *Oceans 2020: Science, Trends, and the Challenge of Sustainability*. Island Press: Washington, D.C., 2002.

Fujita, Rod. *Heal the Ocean: Solutions for Saving the Seas*. New Society Publishers: Gabriola Island, British Columbia, 2003.

Gibson, Ray, Ben Hextall and Alex Rogers. *Photographic Guide to the Sea and Shore Life of Britain and North-West Europe*. Oxford University Press: Oxford 2001.

Gleick, Peter H. *The World's Water, 2006–2007: The Biennial Report on Freshwater Resources*. Island Press: Washington, D.C., 2006.

Granath, Fredrik and Mireille de la Lez. *Vanishing World: The Endangered Arctic*. Abrams: New York, 2007.

Hamilton-Patteson, James. *Seven Tenths: The Sea and its Thresholds*. Faber and Faber: London, 2007.

Helvarg, David. *50 Ways to Save the Ocean*. Inner Ocean Publishing: Makawao, Hawaii, 2006.

Hemingway, Ernest. *The Old Man and the Sea*. Charles Scribner's Sons: New York, 1952.

Holden, Sara. *Planet Ocean: Photo Stories from the "Defending Our Oceans" Voyage*. New Internationalist Publications and Greenpeace International: Oxford and Amsterdam, 2007.

Junger, Sebastian. *The Perfect Storm*. Little, Brown and Company: New York, 1997.

Kunzig, Robert. *Mapping the Deep: The Extraordinary Story of Ocean Science*. Sort of Books: London, 2000

Lanz, Klaus. (Translated by Andew Bird.) *The Greenpeace Book of Water*. David & Charles: Newton Abbott, 1995.

London, Jack. *The Sea-Wolf*. MacMillan: New York, 1904.

Macdougall, Doug. *Frozen Earth: The Once and Future Story of Ice Ages*. University of California Press, 2006.

Melville, Herman. *Moby Dick: or, The Whale*. Harper & Brothers: New York, 1851.

Monsarrat, Nicholas. *The Cruel Sea*. Cassell: London, 1951.

Pearce, Fred. *When the Rivers Run Dry: Water—The Defining Crisis of the Twenty-First Century*. Beacon Press: Boston, 2006.

Philbrick, Nathaniel. *In the Heart of the Sea: The Tragedy of the Whaleship Essex*. Penguin Books: New York, 2001.

Romm, Joseph. *Hell and High Water: Global Warming—the Solution and the Politics—and What We Should Do*. HarperCollins: New York, 2007.

Shiva, Vandana. *Water Wars: Privatization, Pollution, and Profit*. South End Press: Cambridge, Massachusetts, 2002.

Stewart, Iain and Lynch, John. *Earth: The Power of the Planet*. BBC Books: London, 2007

United Nations Educational, Scientific and Cultural Organization (UNESCO) and Berghahn Books. *Water, A Shared Responsibility: The United Nations World Water Development Report 2*: Paris and New York, 2006.

Verne, Jules. *Twenty Thousand Leagues Under the Sea*. (Original translation 1872). Revised edition, translated by William Butcher. Oxford University Press: Oxford, 2001.

Weightman, Gavin. *The Frozen Water Trade*. HarperCollins: London, 2003.

_INDEX

_ACKNOWLEDGMENTS

The sources of data cited in the text are as follows below. **Abbreviations used: ASR** (*Water: A Shared Responsibility*, United Nations World Water Development Report 2); **EB** (Encyclopaedia Britannica); **F&T** (*Facts and Trends: Water*, The World Business Council for Sustainable Development); **IPCC** (Intergovernmental Panel on Climate Change); **IYF** (International Year of Freshwater 2003); **NOAA** (National Oceanic and Atmospheric Administration); **NPA** (Norwegian Polar Institute); **NSIDC** (US National Snow and Ice Data Center); **NWS** (US National Weather Service); **OL** (Ocean Literacy, National Geographic Society); **OP** (Ocean Planet, Smithsonian Institution); **UIMG** (University of Illinois Meteorology Guide); **W&E** (*Water and Ethics: A Historical Perspective*, UNESCO Hydrological Programme); **WE** (Water Encyclopedia, www.waterencyclopedia.com)

Page 19 US National Park Service; **19** WE; **22** ASR; **22** Environment Canada; **22** UN Environment Programme-GRID Arendal (http://maps.grida.no/go/graphic/major_river_basins_of_the_world); **26** WE; **26** World Waterfall Database (www.world-waterfalls.com); **31** ASR; **35** ASR; **35** Institute for Global Environmental Strategies; **35** National Groundwater Association (www.ngwa.org/public/gw_use/faqs.aspx); **35** WE; **37** US National Park Service; **39** EB; **39** UNESCO Water Portal; **39** W&E; **40** EB; **40** W&E; **43** IYF; **43** *Norton Anthology of American Literature*; **43** W&E; **44** ASR; **44** F&T; **44** W&E; **47** Hidden Waters (www.waterwise.org.uk); **49** ASR; **49** F&T; **49** Hidden Waters; **51** ASR; **51** F&T; **51** IYF; **52** *Tall Trees, Tough Men* (Robert E. Pike); **54** ASR; **54** Greenpeace; **54** *The Hidden Freshwater Crisis* (Worldwatch Institute); **57** ASR; **57** EB; **58** ASR; **58** IYF; **58** The Rehydration Project; **60** Aquastat (FAO, United Nations); **60** F&T; **65** F&T; **70** IYF; **70** ASR; **70** Survey of the Scientific Committee on Problems of the Environment for GEO-2000; **70** Vital Water Graphics (UN Environment Programme); **79** NOAA; **79** OL; **80** Goddard Earth Sciences Data and Information Services; **83** NOAA; **83** OL; **85** Ocean World (Texas A&M University); **86** NOAA; **89** EB; **89** NOAA; **91** MarineBio; **91** OL; **91** OP; **92** OP; **95** MarineBio; **95** OP; **97** Coastal Erosion: South Pacific Sea Level and Climate Monitoring Project; **97** WE; **98** Coastal Geology (US National Park Service); **98** WE; **100** EB; **103** EB; **103** Cunard Line; **104** MarineBio; **104** OP; **104** The State of World Fisheries and Aquaculture; **106** The State of World Fisheries and Aquaculture; **109** The Salt Institute; **110** ASR; **113** MarineBio; **113** NOAA; **113** OP; **118** *Quelle eau boirons-nous demain?* (Pierre Hubert and Michèle Marin); **118** US Geologic Service; **118** WE; **121** US Geologic Service; **121** *Vive l'eau* (Jean Matricon); **121** WE; **123** NWS; **123** UIMG; **124** NWS; **124** UIMG; **126** NWS; **126** UIMG; **128** NWS; **128** UIMG; **130** NWS; **130** UIMG; **132** International Organization for Dew Utilization; **132** WE; **135** NWS; **135** UIMG; **139** NWS; **139** UIMG; **140** IPCC; **140** Langley Research Center (NASA); **142** EB; **142** *The Reader's Companion to American History*; **142** US Energy Information Administration; **149** NSIDC; **150** NSIDC; **150** UIMG; **152** NSIDC; **155** NSIDC; **155** WE; **156** NSIDC; **156** WE; **159** NSIDC; **160** Parks Canada; **160** NSIDC; **162** NPI; **164** ASR; **164** NPI; **166** NPI; **168** NSIDC; **170** MarineBio; **170** Ocean World (Texas A&M University); **170** WE; **176** MarineBio; **179** IPCC; **179** MarineBio; **179** NSIDC; **180** NSIDC; **183** IPCC; **183** NPI; **183** National Resources Defense Council; **183** NSIDC; **187** IPCC; **187** NSIDC

_PICTURE CREDITS

The publisher would like to thank the following individuals and photographic libraries for permission to reproduce their material. Every care has been taken to trace copyright holders. However, if we have omitted anyone we apologize and will, if informed, make corrections to any future edition.

l = left, **r** = right, **b** = bottom, **m** = middle, **t** = top, **bl** = bottom left, **br** = bottom right.

Page 4l image100/Corbis; **4r** Ingo Arndt/Getty Images; **5l** J&L Images/Getty Images; **5r** H.Wiesenhofer/Photolink/Getty Images; **6** Datacraft/Getty Images; **12–13** Alan Puzey/Getty Images; **14–15** image100/Corbis; **18** InterNetwork Media/Photolibrary.com; **20–21** Corbis/Photolibrary.com; **22** Galen Rowell/Corbis; **23** Arctic-Images/Getty Images; **24–25** © Yann Arthus-Bertrand/Altitude; **26** JTB Photo/Photolibrary.com; **27** rubberball/Getty Images; **28–29** Willard Clay/Getty Images; **30** Gary Vestal/Getty Images; **32–33** Felix Agel/Photolibrary.com; **34** Demetrio Carrasco/Getty Images; **36–37** Michael Melford/Getty Images; **38** Bruno Barbey/Magnum Photos; **40–41** Jonathan Blair/Corbis; **42–43** John Henry Claude Wilson/Photolibrary.com; **45** Louie Psihoyos/Corbis; **46** Georg Gerster/Panos Pictures; **48** Roy Rainford/Robert Harding; **50–51** Kurobe Dam, Japan – Getty Images; **52–53** J.A. Kraulis/Masterfile Corporation; **54** 2006 China Photos/Getty Images; **55** Jim Wark/Getty Images; **56** Vera Atchou/Getty Images; **59** doug4537/istockphoto; **60** Sabine Scheckel/Getty Images; **63** © Stefano Beggiato/first published in COLORS magazine n° 31-Water; **64** Izzet Keribar/Getty Images; **66–67** Georg Gerster/Panos Pictures; **68** Image Source Black/Getty Images; **70** Yamada Taro/Getty Images; **72** image100/Corbis; **73** image100/Corbis; **74–75** Ingo Arndt/Getty Images; **78** NPA/NASA/Getty Images; **80tl** image100/Corbis; **80tr** Susanna Blåvarg/Photolibrary.com; **80bl** Jean Pierre Lescourret/Explorer, Camera Press London; **80br** Goodshoot/Photolibrary.com; **81** Australia, Whitsunday Islands, Whitehaven beach – Peter Hendrie/Getty Images; **82** Catherine Jouan/Jeanne Rius/Jacana, Camera Press London; **84–85** Bill Alexander/Photolibrary.com; **86** Philip Plisson/Plisson; **87** Masterfile Corporation; **88–89** © Vincent M/Altitude; **89** Pepeira Tom/Photolibrary.com; **90** Blacktail butterflyfish and lyretail anthia – Georgette Douwma/Getty Images; **91** Jeff Rotman/Getty Images; **92t** Ariel Fuchs/Jacana, Camera Press London; **92m** W. Perry Conway/Corbis; **92b** Jeff Hunter/Getty Images; **93** M.Jozon/Jacana, Camera Press London; **94–95** Xavier Desmier/Rapho, Camera Press London; **96** Joao Canziani/Getty Images; **98** Jeremy Woodhouse/Masterfile Corporation; **99** Blondel P/Scope; **100–101** Chad Ehlers/Photolibrary.com; **102** Gilles Martin-Raget/Kos Picture Source; **103** Justin Guariglia/Corbis; **104** Didier Perron/Plisson; **105** Philip Plisson/Plisson; **106–107** © Guido Alberto Rossi/Altitude; **108** A. Guerrier/Scope; **109** Roger Rozencwajq/Photononstop/Photolibrary.com; **110–111** © Yann Arthus-Bertrand/Altitude; **112** Workbook Stock/Jupiterimages; **114–115** J&L Images/Getty Images; **118** David De Lossy/Getty Images; **120** The Smoky Mountains – Karen Kasmauski/Corbis; **122** Design Pics Inc/Photolibrary.com; **123** Cumulonimbus from space – Corbis; **124tl** Brian Stablyk/Getty Images; **124tr** Grant Faint/Getty Images; **124bl** VisionsofAmerica/Joe Sohm/Photolibrary.com; **124br** Mark Newman/Getty Images; **125tl** MedioImages/Getty Images; **125tr** Trapman/Jacana, Camera Press London; **125bl** Eastcott Momatiuk/Photolibrary.com; **125br** Adrian Myers/Getty Images; **126–127** Golden Gate Bridge, San Francisco – Jean Pierre Lescourret/Explorer, Camera Press London; **128** Fancy/Photolibrary.com; **129** Rainman/zefa/Corbis; **130–131** Corbis/Photolibrary.com; **132–133** Karen Massier/istockphoto; **134** Rick Fischer/Masterfile Corporation; **135** John Lund/Getty Images; **136–137** Warren Faidley/OSF/Photolibrary.com; **138–139** John Lund/Getty Images; **140** Wilfried Krecichwost/Photolibrary.com; **141** Pete Turner/Getty Images; **142** Colin Garratt/Milepost 92½/Corbis; **143** Steve Craft/Masterfile Corporation; **144–145** H. Wiesenhofer/Photolink/Getty Images; **148** Wong Adam/Redlink/Corbis; **149** Jean-François Hagenmuller/Jacana, Camera Press London; **150** Hans Strand/Corbis; **151** Ruegner Martin/Photolibrary.com; **152–153** Kenneth G. Libbrecht/SnowCrystals.com; **154** Harvey Lloyd/Getty Images; **156–157** Momatiuk – Eastcott/Corbis; **158** Masterfile Corporation; **159** Steven Allen/Getty Images; **160–161** Chris Coe/Axiom Photographic Agency; **162** Nasa/Photolibrary.com; **165** Antarctic peninsula, Drake Passage – Eastcott Momatiuk/Photolibrary.com; **166–167** Tui De Roy/Getty Images; **168–169** © Steve Bloom/stevebloom.com; **169** Ellesmere Island – Staffan Widstrand/Corbis; **170** Colin Monteath/Hedgehog House/Getty Images; **171** Altrendo/Getty Images; **172–173** Fritz Pölking/Photolibrary.com; **174** Gerard Soury/OSF/Photolibrary.com; **175** Masterfile Corporation; **176** Juniors Bildarchiv/Photolibrary.com; **177t** Daniel Cox/OSF/Photolibrary.com; **177b** Alaskastock/Photolibrary.com; **178** Christophe Boisvieux/Corbis; **179** Arctic-Images/Getty Images; **180–181** Masterfile Corporation; **181** Doug Allan/Getty Images; **182** Markus Renner/Photolibrary.com; **183** F. Jourdan/Hoa – Qui, Camera Press London; **184–185** Candlemas Island, South Sandwich Islands – National Geographic Creative/Getty Images.